Simplexity

Published in association with Éditions Odile Jacob for the purpose of bringing new and innovative books to English-language readers. The goals of Éditions Odile Jacob are to improve our understanding of society, the discussions that shape it, and the scientific discoveries that alter its vision, and thus contribute to and enrich the current debate of ideas.

Simplexity

Simplifying Principles for a Complex World

ALAIN BERTHOZ

Translated by Giselle Weiss

Yale

UNIVERSITY PRESS

New Haven and London

Éditions Odile Jacob

Paris

Translated from *La Simplexité*, by Alain Berthoz, published by Éditions Odile Jacob in 2009. Copyright Odile Jacob, 2009; ISBN 978-2-7381-2169-1.

Yale University Press books may be purchased in quantity for educational, business, or promotional use. For information, please e-mail sales.press@yale.edu (U.S. office) or sales@yaleup.co.uk (U.K. office).

Set in Fournier type by Tseng Information Systems, Inc.
Printed in the United States of America.

Library of Congress Cataloguing-in-Publication Data
Berthoz, A.
[La Simplexité. English]
Simplexity : simplifying principles for a complex world /
Alain Berthoz ; [translated by] Giselle Weiss.
p. cm. — (An Éditions Odile Jacob book)
Includes index.
ISBN 978-0-300-16934-8 (hardback)
1. Science—Miscellanea. 2. Complexity (Philosophy).
3. Simplicity. I. Title.
Q173.B54713 2012
003—dc23   2011015858

A catalogue record for this book is available from the British Library.

This paper meets the requirements of ANSI/NISO Z39.48–1992
(Permanence of Paper).

10 9 8 7 6 5 4 3 2 1

Contents

Preface

How does it happen that a properly endowed natural scientist comes to concern himself with epistemology? Is there no more valuable work in his specialty? I hear many of my colleagues saying, and I sense it from many more, that they feel this way. I cannot share this sentiment. . . . Concepts that have proven useful in ordering things easily achieve such an authority over us that we forget their earthly origins and accept them as unalterable givens. Thus they come to be stamped as "necessities of thought," "a priori givens," etc. The path of scientific advance is often made impassable for a long time through such errors. For that reason, it is by no means an idle game if we become practiced in analyzing the long commonplace concepts and exhibiting those circumstances upon which their justification and usefulness depend, how they have grown up, individually, out of the givens of experience. . . . By this means, their all-too-great authority will be broken.

— *Albert Einstein*

This book invites you to reflect on a new concept: simplexity. I use this term to designate a most remarkable invention of life and one that applies at any number of levels, from molecules to thoughts, individuals to pairs, and ultimately to consciousness and love.

Complexity has become a major buzzword. The economy is complex. Life in megacities is complex. The mechanisms of Alzheimer's disease are complex. Finding the right biofuel to replace gasoline is complex, and so is managing separate families to achieve both the harmonious rearing of children and sexual freedom for parents. We are staggering under the weight of complexity. As if that were not enough, we belong to various social, religious, and political groups and must juggle any number of identities: citizen, neighbor, doctor or bricklayer, tourist, patient, client, or voter. Each of these factors shapes us and imposes on us certain behaviors, norms, customs, and *habitus* (sociologist Pierre Bourdieu's expression for deeply "embodied" ways of doing and thinking) that place us in ever-changing, interlocking social and psychological webs of a complexity unequaled in the history of humankind.

Scientific theories of matter and life must also contend with the complexity of natural processes. No domain is immune. Physics has long been searching for a way out of complexity. Although the discipline is fairly mature, faced with complexity, it must accept the uncertainty relationships that define the very limits of knowledge and admit, for example, that we cannot know both the position and the momentum of a particle.

In an effort to formalize complexity, scholars from all disciplines created an institute devoted to the subject in Santa Fe, New Mexico. Physicist and Nobel Laureate Murray Gell-Mann, the discoverer of quarks, is one of its founders. His book *The Quark and the Jaguar* elegantly summarizes the steps involved in constructing a theory of complex adaptive systems.[1] Everybody is familiar with the metaphor: The flap of a butterfly's wings in Brazil triggers a tornado in Texas. In other words, a very simple law of organization can give rise to complex structures.[2]

The Wonders and Deceptions of Simplicity

Our brains are overwhelmed by the immense quantity of information required to live, act, and understand. In response to the challenges of complexity, ways of simplifying things are proliferating. Intended to keep us all from going mad, these approaches slap on a facade of simplicity in the form of fancy mathematical theories that mask their authors' failure to grasp reality. Motivated by special interests, these mathematical models can lead to calamities, as evident in the recent financial crisis and the failure of the banking system due in part to artificial, completely unrealistic "models" of *"Homo economicus."* Similarly, efforts to facilitate decision making tend to reduce humans to logical processes so as to model them using logico-mathematical theories that simplify real life. But despite the quest for effective solutions, "simple heuristics that make us smart,"[3] the truth is that today we are like the mythical Theseus lost in a labyrinth, without Ariadne's thread to help us find the way. We are made to think that the exit is just at the end of the corridor, but the corridor leads nowhere. Lost in the genuine complexity of the world, aware of the ineptitude of formal models, we are easy prey for fundamentalist beliefs and obscurantism.

This need to simplify touches everything. The quest is evident in all areas of social and political life, of medicine, science, technology, and day-to-day existence. The complexity of electronic gadgets is disguised by their ease of handling. It takes huge software programs to make computers user-friendly. Tax forms and medical protocols are being simplified, as are administrative documents. Criminal proceedings are being simplified to accelerate them. We can now vote electronically and have a simple choice between candidates we see debating on a television screen. People's lives are being simplified by creating supermarkets where they can find all the product "solutions" they need. Engineers are trying to simplify the design of "light pipes," [4] and chemists have uncovered simplifying principles for enzymatic and kinetic reactions.[5] The result of this frenzy of simplification is accrued complexity. The easier computers are to use, the more bloated the software. Simplification costs.

These days, there is a tendency to confuse modernity and simplicity. Oversatiated with the pervasiveness and exuberance of baroque art, the whims of classical architecture, and the extravagant refinements of suits and dresses, the twentieth century welcomed a reductionist movement in favor of the simplest shapes and materials. Typified by the influential Bauhaus school, the movement came to dominate industry and design. Fortunately, we are now seeing a countermovement, and clothing designers, for example, appear to have rediscovered the joy of playing — in the musical sense of the term — with forms and colors, textures and rhythms, and with the flow and folds of material.

The Originality of Life

Let me define more precisely what I mean by simplexity.[6] I did not invent the concept. The word *simplexity* has been used by geologists since the 1950s, and it is common in the fields of commerce, design, and decoration. Nonetheless, this usage is of limited interest to us, for it is often intended to be synonymous with *simplicity*. For me, simplexity means something else. First and foremost, it is a property of life. In this book, I will examine the concept of simplexity and its significance in an effort to better understand what makes life unique. Simplexity is not simplicity. It is fundamentally linked with complexity, with which it shares common roots. As Gell-Mann writes, "Simplicity refers to the absence (or near absence) of complexity. Whereas the former word is derived from an expression meaning 'once folded,' the latter comes from an expression meaning 'braided together.'" [7]

Some complexity theorists have striven to identify what distinguishes living organisms from inert matter. According to one of these theorists, physicist and mathematician Nicola Bellomo, "Although living systems obey the laws of physics and chemistry, the notion of function or purpose differentiates biology from other natural sciences. More important, what really distinguishes biology from physics are survival and reproduction, and the concomitant of function."[8] Consequently, he proposes a mathematical theory of interactions between "a large number of interacting entities which will be called active particles, or occasionally agents, and which are generally organized in different interacting populations." As interesting as this concept of action is, it has its limits. It does not really deal with what is most original about action in living organisms. Nowhere is there any mention of the idea that life has found solutions to simplify complexity. Nor does it evoke the utterly remarkable ability of living creatures to create borders delimiting closed spaces, such as the cell and the body itself. These solutions are indeed simplifying principles that reduce the number or the complexity of processes. They make it possible to rapidly analyze information or situations, taking into account past experience and anticipating the future — which helps to grasp intention — all the while respecting the complexity of reality. In my view, the ensemble of these solutions is simplexity. Simplex solutions are not just ways of reformulating or summarizing a problem. Put another way, they enable actions that are more elegant, faster, and more efficient. They also give priority to the senses, even if it means making a detour.

Simplexity is complexity decoded, so to speak, because it is based on a rich combination of simple rules.[9] To borrow a phrase from the mathematician and philosopher Gottfried Leibniz, about the best of all possible worlds being one that combines the greatest variety of phenomena with the simplest laws, it is "complicated simplicity."[10] The music of French composers Pierre Boulez and Pascal Dusapin is modern; you do not have to like it, but it is simplex. Also simplex is a Bach fugue, which begins with several notes and evolves slowly toward soaring whorls of combined sounds. It seems complex, but the notes actually follow rigorous laws. Another example is found in the great Russian liturgies, where the polyphonies give the illusion of great simplicity through a skilled arrangement of rhythms and sonorous spaces. These overlap and intertwine in a dance that seems to be a solo in the sublime way it harmonizes the multiple activities going on in the brain.

Simplifying in a complex world is never simple. In particular, it requires us to choose, refuse, connect, and imagine. I have said elsewhere that the basis of our thoughts, from the development of our highest cognitive functions, even the most abstract, lies in action, and that our brain evolved to anticipate the consequences of an action, projecting onto the world its preperceptions, hypotheses, and interpretative schemas. The originality of life is precisely that it found solutions to resolve the problem of complexity by mechanisms that are not always simple, but simplex. It is possible to have the impression that complexity is reducible to a mouse click on a computer, that the world really is within reach of a Google Web page, that to solve the major psychiatric disorders we have simply to discover their genes. Such an approach suffices for operating a washing machine, a computer, or a ticket vending machine at the train station. But it is useless faced with the genuine problem of how to integrate the multiple complexities thrown up by our social, material, and natural environment. Simplexity theory complements complexity theory. In some way, it contains complexity. Laying the foundation for a biologically founded theory of simplexity is the modest aim of this book.

Acknowledgments

First of all, my thanks to Odile Jacob and Bernard Gotlieb, who were receptive to the idea of a book on simplexity. Over the years, both have provided me with countless opportunities for intellectual and epistemological debate and encouraged me to probe the edges of disciplines in which I work. I would also like to thank my editor, Marie-Lorraine Colas, for her wise council and her work on the text and ideas, and the entire editorial team at Éditions Odile Jacob for their endless patience with an author who cannot stop fiddling with his books.

Thanks to Vincent Laborey, who first encouraged me to write. My endless gratitude to my wife, Maya Berthoz, for her apropos and fruitful suggestions regarding the book. Moments of doubt are an inescapable part of an endeavor like this one, and no remedy is more effective than enduring tenderness. Thank you to Robert Baker, Philippe Descola, Claude Debru, Jacques Droulez, Benoît Girard, Olivier Faugeras, Jean-Paul Laumond, Giuseppe Longo, Jean-Pierre Nadal, Jean Petitot, Béatrice Picon-Valin, Alain Prochiantz, Jean-Jackques Slotine, Brian Stock, and other colleagues who were willing either to reread the text or to provide me with material in writing it. My appreciation also to my young co-workers in the laboratory who helped me with the content, figures, and references.

Daniel Bennequin was unstinting in his support for the sections regarding the notion of simplexity in mathematics. I could not have written those parts (and many others) without him. It is an indescribable privilege for me to have the benefit of his immense knowledge, which goes far beyond the bounds of mathematics.

Thank you, more than I can say, to France Maloumian, our laboratory's infographics artist, who once again not only captured the essential in the illustrations but who researched, compiled, and formatted the source information that accompanies them. Her help in organizing the book was invaluable. My thanks also to Hélène Leniston for her work on the text.

I am grateful to the Collège de France and to Centre National de la Recherche Scientifique, which both continue to offer me the unique environment a researcher needs to freely test new ideas. Thank you to everyone in the laboratory and on the administrative staff for contributing to the atmosphere of science and trust that allows me time to think.

My heartfelt thanks also to Giselle Weiss, who did the English translation. Her ability to find the words, style, and organization that best suit what I want to say is remarkable.

I am equally indebted to the entire team at Yale University Press for their very thorough and professional work on this edition of the book. Their elegant mixture of rigor and flexibility is an immeasurable kindness to an author who requires both guidance and freedom.

I dedicate this book to my six grandchildren and to those who will come. To all of them I say, "Remember to dare."

Simplexity

Part I. REMEMBER TO DARE

In certain circumstances supreme boldness becomes supreme *prudence*.

— *Carl von Clausewitz*

1. Making the Complex Simplex

Simplifying principles gives hope that the behavior of seemingly incomprehensible biological networks will eventually be deciphered. I have emphasized simplicity in biology to encourage the point of view that general principles can be discovered. Without such principles, it is difficult to imagine how we might ever make sense of biology on the level of an entire cell, tissue, or organism.

— *Uri Alon*

Why propose the neologism *simplexity* to describe the properties of life when the term *simplicity* already exists? It is more than just a play on words. The word connotes the remarkable fact that biological devices, or processes, appeared in the course of evolution to allow animals and people to survive on our planet. Given the complexity of natural processes, the developing and growing brain must find solutions based on simplifying principles. These solutions make it possible to process complex situations very rapidly, elegantly, and efficiently, taking past experience into account and anticipating the future. They also enable us—by means of a fundamental principle of "intersubjectivity"—to understand the intentions of others. They do not make reality any less complex. They may involve *detours*, an apparent complexity, by presenting problems in a novel way, changing reference frames, points of view, and so forth. Contrary to what

3

we might think, simplifying is not simple, for it requires us especially to refuse, inhibit, choose, connect, and imagine. Some solutions devised by life are universally valid for all species of animals, including humans. This is true of the senses. But each organism also comes up with solutions as a function of its *Umwelt* (a concept we have examined in the book *Neurobiology of "Umwelt"*);[1] that is, the organism's own relation with the environment based on its position in the phylogenetic tree. Even the complex set of genes, their expressions, and probably also their associated environmental influences (epigenesis) are organized according to principles that both enable simplification and promote diversity. This is how, for example, four families of genes — only four! — control the organization of the body into segments, yet the anatomy of the cranial nerves still differs depending on the species (see fig. 1). Let us take a little closer look at genetics and simplexity.

Patterns and Small Worlds

Uri Alon, an Israeli physicist who is looking to uncover the general principles underpinning biological circuits and networks, has this to say on simplicity:

> "Complex" is perhaps the most common adjective used to describe biological phenomena. In every cell, complex networks of interactions occur between thousands of metabolites, proteins, and DNA. Every interaction is itself a complex dance between exquisitely shaped proteins, designed to interface with each other if the conditions are right. And every protein looks like tangled strands of spaghetti festooned with atomic appendages. So where is the simplicity? . . . The point I wish to make is not that biology is simple, but that biological networks of interactions are simpler than they might have been.

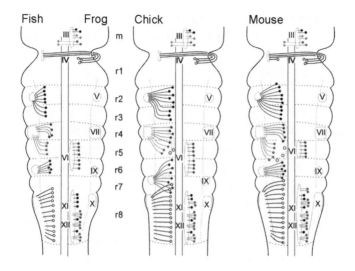

Figure 1. The simplex anatomical organization of the cranial nerves. The four HOX genes that determine the segmentary localization of the parts of the body also determine the distribution of the nerves along the spinal cord (following the "rhombomeres," numbered r1 to r8). This simplification (only four genes) enables enormous diversity in the distribution, depending on the species, as here in the example of cranial nerves. [Adapted from E. Gilland and R. Baker, "Evolutionary Patterns of Cranial Nerve Efferent Nuclei in Vertebrates," *Brain, Behaviour, and Evolution* 66 (2005), 234–254.]

There seems to be a degree of simplicity in several aspects of these networks, which is intriguing given that cells evolved to survive, and not for scientists to understand.[2]

Despite the incredible quantity of patterns of interaction between genes, the neuronal networks they give rise to are constructed from a very small number of what Alon calls "motifs." For example, the bacterium *Escherichia coli* contains a motif that responds to stress by producing flagella that resemble whips. These enable the bacterium to swim toward more-hospitable areas. This same motif also turns up in hundreds of other systems, including very complex organisms. These specific patterns appear to be related to the need to develop robust structures.

Replication of identical or similar motifs that fulfill specific, important functions is a good example of what I call simplexity. Identical or similar motifs are used throughout the living world to minimize energy, reduce entropy, and even to transmit information faster. Simplexity also appears in molecular assemblies for which, sometimes, the basic principles consist simply in the ability of the constituents to produce reciprocal patterns of excitation and inhibition as a function of chemical concentration. In the same text cited here, Alon comments, "Such models seem to capture the essential dynamics of protein circuits, while being, in a sense, insulated from most of the complexity of the proteins themselves."[3] It stands to reason that modeling can be a powerful tool for studying simplifying principles to describe what many biology researchers call "small worlds." Other examples of efforts to identify solutions that simplify a given process are to be found in different areas of biology, such as immunology, but also in disciplines that study the molecular basis of behavior.[4]

Tools for Life

For the purpose of argument, let me suggest a preliminary list of basic characteristics of life that I believe rely on simplex properties that constitute tools for life:

Modularity. Separation of function is an essential feature of life. It is well known that there are several visual pathways, for example, for orientation (via the colliculus), emotion (via the amygdala), identification (via the so-called ventral pathway), and localization and context definition (via the dorsal pathway). Different types of memory—explicit, implicit, procedural, and so forth—have separate neuronal networks. We also have separate neuronal loops for the control of movement between the basal ganglia, the thalamus, and the cortex, for eye movements, limb movements, memory, and emotion. A number of descending and ascending pathways between the brain and the spinal cord subserve different functions. Although we must refrain from an excessively localized theory of brain processes, it is still true that different areas in the brain process specific aspects of perception, action, memory, and emotion, each belonging to ever-changing networks that interact dynamically through such mechanisms as direct action and synchronization of oscillations. Likewise, there are separate and coordinated mechanisms for automatic behavior and cortically controlled action. This diversity is a feature of simplicity. More generally, modularity (the equivalent of our "division of work" in society) is a fundamental property of living organisms.

Speed is another fundamental property of life and of the brain. Very few animals can manage without it; the sloth is an exception. Speed demands elegant solutions, not necessarily simple, but efficient. It calls for anticipating the consequences of action, which is indispensable in capturing prey or escaping from a predator. Thus, the praying mantis, fearsome devourer, moves on the bee that it wishes to trap in 60 milliseconds. In molecular assemblies, as well as in the higher activities of the nervous system, functions are distinguished by their temporal dimension. For instance, some molecular mechanisms work rapidly and others slowly. Similarly, in motor control, but also in perception, "tonic" (slow, sustained) systems are distinct from "phasic" (rapid, transitory) systems. Sensorimotor functions are divided into specialized modules that cooperate (see fig. 2). In humans, it takes between 75 and 100 milliseconds for the sight of a snake to trigger fear. Emotion pathways are also divided into fast- and slow-

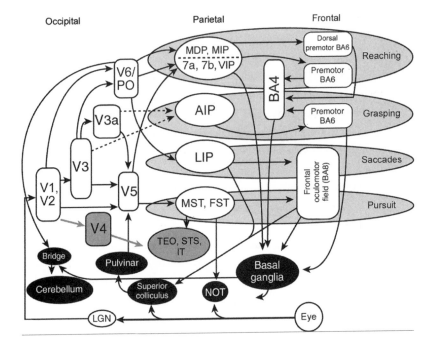

Figure 2. Modularity: different neuronal networks supporting a repertoire of actions. This schematic drawing shows the modularity of the systems for reaching for an object, grasping it, redirecting the eyes by saccades, and visually pursuing a moving target. LGN indicates the lateral geniculate nucleus, the center of the visual thalamus. V1 to V6/PO indicate the different visual (occipital) areas; MDP, MIP, 7a, 7b, VIP, AIP, and LIP are regions of the parietal cortex; MST, FST, TEO, STS, and IT designate the areas of the temporal cortex. The areas labeled BA correspond to the frontal and prefrontal cortex, according to Brodmann's nomenclature. [From J. Atkinson, *The Developing Visual Brain* (Oxford: Oxford University Press, 2002). Reprinted by permission of Oxford University Press.]

responding: In the presence of a snake the fast amygdala pathway triggers fear and reaction in less than 100 milliseconds, while slower cortical pathways analyze the object of fear and eventually modify the fear response. The anticipatory motor action produced by an obstacle that could trip you activates at 100 milliseconds. This kind of speed is also found in more-complex cognitive acts, such as decision making, which sometimes is accomplished in a flash. An airplane pilot has very little time to avoid a catastrophe. A bus driver who takes his eyes off the road for more than a second is asking for trouble. Playing a violin piece by Paganini or Mozart requires a dazzling dexterity—the lightning-quick translation of thought into action—whose only equal is the mental dexterity of the composer. To enable this execution, life has devised a problem-solving method similar to that formulated by seventeenth-century French philosopher and mathematician René Descartes: Break complicated problems down into simpler subproblems, thanks to specialized modules, though, of course, everything must be put back together again. I call this the simplexity detour.[5]

Reliability. To avoid errors, the neuronal mechanisms of the brain and its parts must be highly reliable. But reliability is no friend of complexity, especially when it involves living organisms at several levels, from the molecular up to the cognitive. Novel solutions for increasing reliability do exist, however, among them redundancy, noise (paradoxically), cooperation between inhibition and excitation, and coupled oscillators. As we will see below, it is now also believed that the brain uses probabilistic processes to cope with uncertainty. These solutions may be endowed with properties such as "contraction" to aid in coordinating and stabilizing the thousands of neural loops involved in brain operations.

Flexibility, vicariance, and adaptation to change. An organism must be able to resolve a problem, perceive, capture, decide, or act in several different ways (vicariance) to adapt to context, compensate for deficits, and face new situations. But to have a repertoire of solutions at the ready, it must not get bogged down in complexity. Suppose that I have to be at the Panthéon at Porte Maillot in Paris at a certain hour. The simplest solution is to take the bus, which is direct. But traffic is jammed all along the way, and I will be late. So instead I take the regional rail, which requires me to change at Châtelet, a complicated station where everybody gets lost. Amazingly, I arrive on time! My solution introduced a bit of complexity, but it simplified my life: I was not late. Another example, on a different level: Say I have a portfolio of stocks and the market starts to plummet. The simple solution is to sell my shares quickly. The simplex solution, both simpler and more complex because it requires a mental detour, consists in doing nothing while waiting for the market to go back up. This involves reasoning, betting on the future. Compared with simplicity, simplexity includes a tension, sometimes an opposition between simple and complex, that is characteristic of life.

Memory. Present action relies on the memory of past experience to predict the future consequences of action. Connecting the past with the restless present calls for common simplifying principles between the memory of the past and the way in which anticipation is coded in the brain. The multiplicity of mechanisms of memory (explicit, implicit, episodic, verbal, iconic, affective) bring us back to modularity. But memory is not only a property of higher brain mechanisms. All levels of the brain have various mechanisms of memory. Even motoneurons, for example (the neurons located in the spinal cord that activate muscles), may enter a mode called the plateau potential, which is characterized by sustained activity.

Generalization is another very important property of simplex systems. An example is what we call motor equivalence. You can shift your gaze by moving your eyes, but also by a combined movement of the eyes and head, even your entire body. To do that, the movement must be encoded, programmed in a fairly general way so it can be executed by any of the segments of the body, whatever its complexity. I propose that, to ensure this generality, the shift of gaze is encoded in the form of speed, with dynamic-memory mechanisms,[6] which makes it possible to integrate the speed signal and transform it into a position signal by processing it in the internal models of the specific effector.[7] Similarly, the word *love* can be written with a finger, a hand, or even a foot — as people do who cannot use their hands — but you can also do it while running on the beach. This suggests that the geometry of movement is determined in a very general way. We will allude to this property further on. This question of generalization is central to current thinking about rehabilitation of motor function in neurological patients with brain lesions. Robots and machines for re-educating these patients exist, but when they have been trained on the machine, patients have trouble in transferring, that is, generalizing, their relearned capacities to their everyday lives. Many other examples of the capacity to generalize can be found, for instance, in perception and language. Naturally, this list of properties is far from exhaustive, and we will come back to them, in particular regarding the importance of movement and action, which are the foundation of thought, as nicely articulated by neuroscientist Rodolfo Llinás.[8]

2. Sketching a Theory of Simplexity

> It is not because behavior is simpler that it is preferred; on the
> contrary, it is because it is preferred that we find it simpler. . . .
> For the most part preferred behavior is the simplest and most
> economical *with respect to the task in which the organism finds
> itself engaged;* and its fundamental forms of activity and the
> character of its possible action are presupposed in the definition
> of the structures which will be the simplest *for it,* preferred in it.
> — *Maurice Merleau-Ponty*

I would like to try to sketch out a theory of simplexity. A sketch is not
a final drawing; it is the expression of an intention, an idea, imprecise
and indecisive, the bearer of its own evolution. It is a question that
hints at its response, a kind of free association. Let me suggest that a
simplex process is one governed by several *principles,* implemented
successively or in parallel. My list of principles is intended to define
a framework, incomplete and open to discussion, whose aim is to
delimit the concept of simplexity. To avoid any misunderstanding,
let me emphasize that I do not deny the fundamental value of complexity theories. I would just like to throw open a window and start a
conversation, very humbly, in reference to the mathematicians who
envisaged non-Euclidean geometries, or to Leonardo da Vinci, who
abandoned painter Leon Battista Alberti's overly strict approach to
perspective, and, more recently, to psychologist and Nobel Laure-

ate Daniel Kahneman, who rethought the cognitive basis of economic theory and challenged the rational nature of human decision making.

Inhibition and the Principle of Refusal

Inhibition is a remarkable functional property of living organisms and of the human brain. It is one of the greatest discoveries of evolution. It enables competition and, consequently, decision making, plasticity (flexibility), and stability. It is used in the brain to enhance speed, to select from among the complex constituents that make up any phenomenon, act, or situation, whether concerning our relationship to the environment or the mechanisms of our thought processes. All the major centers of the brain (cerebellum, basal ganglia, prefrontal cortex) that are involved in coordinating movement, choosing one action among many, predicting the future, or deciding have inhibitory mechanisms at their disposal. For example, the development of the prefrontal cortex allowed humans not to be slaves to lived reality, the flow of events, the world, but rather to keep their distance from reality, to change their point of view. Our executive functions[1] give us the capacity to inhibit primitive cognitive strategies[2] or innate reflexes that kick in too quickly. One might say that to think is to inhibit and disinhibit; to create is to inhibit automatic or learned solutions; to act is to inhibit all the actions that we do not take. The refusal to lose ourselves in complexity is an attitude, an intellectual stance that enables reexamination, a bit like the "bracketing" (mentally setting aside presumptions) so dear to the German phenomenologist and mathematician Edmund Husserl.

The Principle of Specialization and Selection: *Umwelt*

The most striking example of my second principle is the repertoire of sensory cues used by each different species of animal. One species scans the world only for cues important for its survival. The tick knows about the world only through the smell of butyric acid and the heat that signal the presence of a living animal whose blood can be sucked! Most animals act according to their *Umwelt;* they sense only those aspects of the world that are relevant for their survival. We can generalize this idea and apply it to cognitive function and decision making in general. Deciding involves selecting from the information around us whatever is pertinent to the goal of action. It is a principle of parsimony, identical to that at work in the art of war, politics, and reasoning, or that expressed in popular wisdom in the form of proverbs and sayings such as "He who grasps at too much loses all" or "A bird in the hand is worth two in the bush."

This selection is not only induced during a stimulus-response process. It is intrinsic to adopting a perspective, whereby a living, self-organizing, autonomous organism projects its intentions and hypotheses onto the world. In this act, our brain is more a comparator and an emulator than a simple information processor. It uses numerous attentional mechanisms that I will address in a later chapter. This preliminary filtering of information from the world is itself the result of a specialization. As we already pointed out in the previous chapter, modularity is essential: The brain is formed of centers dedicated to certain kinds of processing — vision, the body, memory, language, and emotion.[3] The uniqueness of humans is that, to be able to create worlds, at least to some extent they have the illusion that they can escape their *Umwelt*!

The Principle of Probabilistic Anticipation

The third principle is anticipation based on memory. This double strategy, both prospective and retrospective, situates the present in the dynamic flow of a changing universe. It enables comparison of sensory data with the results of past action and prediction of the consequences of ongoing action. Recent data show this double control at the level of the thalamus, which processes sensory information.[4] Anticipation founded on memory requires the brain to operate in a world full of uncertainties and therefore probabilistic functioning, which is not simple. We can only estimate the speed of our body in space, tomorrow's weather, or the behavior of the stock market. It is no surprise, then, that roboticists rely so heavily on the Kalman filter, a probabilistic mathematical tool. For our purpose, what is important is that prediction is always probabilistic. In fact, today, researchers in psychology and neuroscience use models derived from a theorem formulated by Thomas Bayes, an eighteenth-century British mathematician and Presbyterian minister, who drew a connection between past and future probabilities (see fig. 3). Of course, other probability theories could and will be proposed, but for now, Bayesian inference is particularly useful in modeling a variety of human processes.

Here is a brief synopsis of Bayesian theory. To take action, the brain must make some hypotheses. It must decide what the probability is that its hypotheses are correct, on the basis of the information available, its memory of the past, and its predictions for the future. Think of how you make up your mind about what the weather will be today or tomorrow. Bayes's theorem tells you what to do. If $P(H|D)$ is the probability that a hypothesis is true given the current sensory data, then $P(H|D) = P(D|H) \times P(H) / P(D)$, where $P(D|H)$ is the likelihood, that is, the probability, of these data if the hypothesis is true, $P(H)$ being the a priori probability that the hypothesis is true and $P(D)$ the probability of the data.

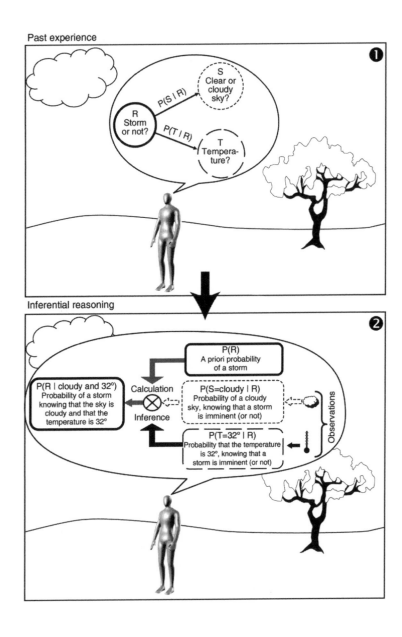

Despite Albert Einstein's resistance to the idea at one time, physicists now believe that the universe is governed by the laws of quantum physics, where indeterminacy reigns. An example is the uncertainty principle, published in 1927 by German theoretical physicist Werner Heisenberg, which describes the extent to which it is possible to simultaneously know the position and speed of a particle. In the same way, simplexity seems to me to resolve complex problems by rejecting dry determinism in favor of probability — chance — the idea that order can emerge from disorder. My nephew married a young Indian woman, and I was given the honorific *mama* ("uncle") because, for a couple in India, it is a sort of godfather. I asked what my role should be. My nephew explained to me that it is not only to ensure the happiness of the couple but to stir it up a little, in other words, to enrich it with the variety so often cited by neuroscientist Jean-Pierre Changeux as a predilection of the brain.[5]

The advantage of this use of probabilistic approaches is a certain margin of freedom. Simplex order is not fascist but democratic, not hierarchical but heteroarchical (containing both hierarchical mechanisms and parallel ones). Like a Bach fugue, simplexity leaves in-

---

(Opposite) Figure 3. Predicting a storm (schematic representation of probabilistic reasoning). Will it rain? To answer this question, you need the information indicated in the first image. The prior information defines the dependencies among the data — such as the condition of the sky (S) or the temperature (T) — and whether a storm appears imminent (R). These dependencies translate into the probabilistic laws P(S | R) and P(T | R), called likelihoods. Since in this example R does not depend on any other variable, P(R) is the a priori probability of a storm, that is, in the absence of information. The second image shows how Bayes's law enables one to use this knowledge to calculate the a posteriori probability of a storm, by taking into account observations regarding the sky and information about the temperature.

numerable variations to the imagination. It is key to innovation. Although I cannot prove it, I have the impression that simplexity is as good as (if not better than) complexity at giving rise to innovation because of its unique spatial and temporal organization.

The Detour Principle

The fourth principle of simplexity is the detour principle, through an accessory complexity. This idea of detour is fundamental, and it must be accepted for what it is, as when *The Michelin Guide* suggests that a certain site or town "is worth a detour." Let us take an example. Imagine that a roboticist wishes to control the position of a robot whose task is to seize objects without knowing their dynamic properties, in a complex environment that is constantly changing, say, catching a paper airplane in flight in windy conditions. The robot must solve "nonlinear" problems. To do that, it replaces the simple variable it wishes to control (position) with a more-complex mix of variables, including position, speed, and acceleration — what we call composite variables.[6] Paradoxically, expressing the problem in terms of composite variables simplifies it. If a system exhibits complex behavior that normally can only be represented by "third-order" equations, using composite variables gives a first-order system that is simpler to calculate and whose dynamic behavior is easier to predict. Similarly, if the speed of a system's changes in position varies in a nonlinear fashion, employing composite variables makes it possible to perform the required calculations in a world of linear speeds that is much simpler. In both cases, there is indeed a detour that appears complex (using composite variables) but that actually results in simpler and more-efficient control of the system. This corresponds to my definition of simplexity.

Another example is computer simulation of an airplane, such as the giant Airbus A380. The simulation it was subjected to before being built required extraordinarily sophisticated algorithms and graphic technologies, but the result of the detour was a greatly simplified process. Or take robotic surgery: Today, a surgeon no longer needs to operate directly on an organ; rather, he manipulates an image of the organ transmitted by a camera placed in the body of the patient. A motorized sensor located at the end of the instrument measures pressure and feeds this "force" back to the hand of the surgeon, giving him the impression that he is touching the organ. This virtual-reality approach is becoming increasingly common in operations such as abdominal endoscopy. It calls for a high degree of dexterity and very specialized training, but it facilitates the task by guiding the surgeon's gestures and by enlarging or decreasing the size (scale) of the image of the organ to be resected or repaired. In the end, this detour makes the surgeon's work easier by incorporating flexibility, memory, and the possibility of consulting a data bank of images and of making his movements more precise. According to my theory of simplexity, living organisms also possess numerous mechanisms that, by means of detours, facilitate the solution of nonlinear problems. Moreover, it is precisely the nonlinear nature of the detour that is key. Note that the detour is not the only elegant solution cooked up by life. Shortcuts exist as well. We will return to this subject in chapter 5.

The Principle of Cooperation and Redundancy

The cost of specialization and selection (our second principle) is the duplication and creation of a substantial amount of information. However, selection reduces the number of available solutions. In such a context, having several values for the same variable to mitigate the risk of error is extremely useful. By cooperation I mean, for example, the fact that we often have several possible ways to evaluate important aspects of the relationship between the state of our bodies and the world. Suppose that you wish to evaluate how fast you are moving. The proposition is complex, because the visual environment is often jittery with motion (a train, clouds, wind). To accomplish your task, your brain must use two independent measures: a specialized sensor as well as a combination of information provided by various other sensors. For the brain to accept the result, the two estimations must be coherent.[7] It is therefore more than just redundancy. We also know that the speed of the head is evaluated twice, by the vestibular system (the three semicircular canals and the two otolith organs located in the inner ear that detect acceleration of the head), and by vision in connection with the measurement of the eye movements.

Cooperation and redundancy are more than just a combination of sensory detectors. They have other fields of application. For example, evolution has provided mechanisms that enable you to envisage the city you are in either egocentrically — that is, first person — based on the route you are following (local perspective), or allocentrically — that is, by imagining a map of the city from a cartographic, global perspective — which has the advantage of permitting you to carry out independent mental operations, such as comparing distances or searching for alternative routes (see fig. 4).

These two perspectives are complementary and constitute a form of simplexity. By employing the detour of the two perspectives and working with them either in parallel or simultaneously, you can simplify your movement in a city, the subway, or a forest. This detour also enables the CEO of a company or a military strategist to simplify the complexity of a construction site or a battlefield. We will touch on the neural bases of the cognitive strategies underpinning these different ways of treating space later on. For now, suffice it to say that perspective helps us to make decisions. Decisions are fundamentally simplex in that they offer an alternative to a complex reality: A surgeon either operates on a patient or releases her; a judge condemns a defendant or acquits him; a stockholder sells her shares or waits

(A) "Route" perspective (egocentric)

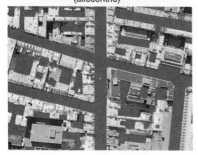

(B) Survey or map perspective (allocentric)

Figure 4. Two distinct cognitive strategies for remembering a route traveled in a city. (A) An image of a town that corresponds to the "lived experience" of an outing: The point of view is egocentric. (B) A survey or cartographic image. Although it involves a degree of subjectivity, this allocentric point of view enables mental operations that are independent of the subject. These two perspectives correspond to two brain mechanisms for simulating and memorizing trajectories, but there are others. The many cognitive strategies for processing space constitute a modularity that allows the brain to choose a mode appropriate for the task or the context. [The virtual-reality reconstruction is courtesy of Archivideo.]

for the market to recover; the police chief forbids a demonstration or gives it a green light; a car manufacturer increases the number of models or chooses to specialize. These choices are made depending on context, rules, points of view, and previous decisions, which serve as "frames of reference" (see chapter 10).

The Principle of Meaning

Thanks to all these properties, for living organisms, simplexity is what gives meaning to simplification, insofar as simplex solutions are motivated by intentions, goals, or functions.[8] As I showed in my book *The Brain's Sense of Movement*, the basis of meaning is in the act itself: Meaning cannot be superimposed on life; it *is* life. To my mind, the concept of simplexity includes the idea of meaning. Elaborating a theory of simplexity thus also entails elaborating a theory of meaning by redefining the term to incorporate the intended or desired act as fundamental. At this stage, the principles I have sketched here are an invitation to debate the idea of simplexity. I do not pretend that they are exhaustive, nor is their formulation definitive. The following chapters will give examples of simplex processes.

3. Gaze and Empathy

The combination of *re* and *gard* [French for *gaze*] is richly con-
notative. More than just the fact of capturing a view, an image,
it evokes the reconsideration and revival of something that was
looked at and that, at every reprise, asks to be developed further. Moreover, *regard* also implies attitude; it compels the person who is doing the gazing to a deeper engagement.

— *François Cheng*

Throughout evolution, solutions have been devised to permit living
organisms to act rapidly and efficiently. My hypothesis is that incredibly numerous and varied solutions — the diversity of life — are
found in very different organisms. Simplexity responds to the same
rules as language or culture: It encompasses both diversity and universality, as is evident in the opposing ideas of American linguist
Noam Chomsky (who stresses the universality of grammar)[1] and his
French contemporary Claude Hagège (who emphasizes the diversity
of language).[2] The problem is present as well in the work of the late
anthropologist Claude Lévi-Strauss, who based his structural an-
thropology on the detailed examination of various concrete forms
of social organization in order to extract from them generalizable
formal structures.

In fact, this diversity-universality duality is a characteristic of
life. It is found in genomics, where recent discoveries have helped

us understand how chemical mechanisms of metabolism in similarly organized genomes induce variations that manifest as diverse phenotypes; how two plants that have the same genome can have completely different geometries, that is, different arrangements of branches and shapes of leaves and flowers. Today, we are also beginning to recognize the importance of epigenetic factors, either special genes or environmental or behavioral factors, which influence the expression of genes from the very outset of an infant's life.

In describing simplex solutions, for lack of anything better, I will employ terms related to simplifying laws and principles. In reality, it is neither laws nor principles that are at issue, but solutions, processes, architectures, and even mechanical agents. Evolution seems to have made use of everything available, borrowing nonobvious pathways, using all the tools that physics and chemistry have to offer with the sole aim of simplifying. Since I am a physiologist, I use examples from the most integrated level of functioning of the organism.

The Repertoire of Gaze Movements

A first illustration of simplexity, borrowed from life, is provided by gaze — more precisely, by gaze movements. The eye appeared very early in evolution. It is a critical invention, for it enables the prime faculty of life, movement, whether it be to capture prey or to escape a predator. Just consider the sea squirt: half plant, half animal, endowed with a tiny brain that controls a mini motor apparatus as well as an eye and a vestibular sensor.[3] When the organism has exhausted the food supply in its fixed environment, the motor apparatus allows it to get around, the eye to see its surroundings, and the vestibular sensor to stabilize its position and orientation in the water, thanks to a mobile frame of reference, as it searches through currents for a new site in the uncertain universe of the ocean. The eye is mobile, which implies a gaze mechanism, for its vision must be able to sweep the space around the animal.

Gaze is an extraordinary invention and a fundamental element of the simplexity of life, not least because it enables anticipation and adapts visual exploration to intended action. Gaze reduces the complexity of analysis of the visual world by focusing attention; it stabilizes images on the retina. Finally, it maintains objects on the fovea—the area of the retina most sensitive to detail—through the mechanism of ocular pursuit, that is, the capacity to slowly follow a moving object. The study of these mechanisms is a delight: The more we discover about them, the more we are amazed by their variety.

Yet, all it takes to render these properties useless is a slight shift of the image of the world on the retina. Moreover, we need to perceive moving objects, situated at different distances, and to isolate the object of interest in this moving world. Although it is very complex, evolution has several solutions to the problem of moving the eye to immobilize an image on the retina. A critical component of the solution is specialization: We possess a very rich repertoire of eye movements, each of which has a precise function (see fig. 5).

Reflexes, either of vestibular or visual origin, stabilize the image of the world on the retina; successions of small, quick movements, called *saccades*, which we all make when we visually look at the environment, enable us to explore the visual world; *ocular pursuit* (mentioned above) makes it possible to follow a moving object in a continuous manner. In addition, other mechanisms (not shown) allow movements of *convergence and accommodation*, which help to adjust gaze to absolute distance. The mechanisms that permit the brain to simplify the coordination of these movements are too numerous and remarkable to cover in detail here. What is important is that each of these subsystems is controlled by a specialized neuronal mechanism

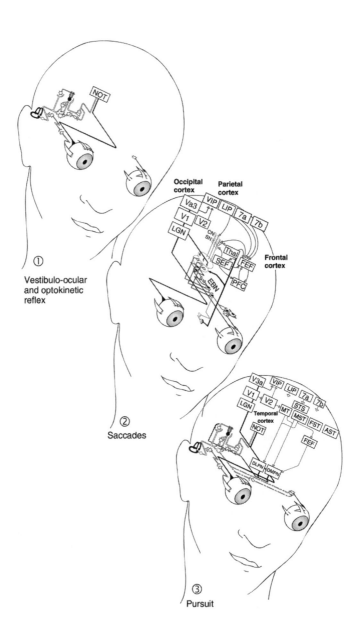

① Vestibulo-ocular and optokinetic reflex

② Saccades

③ Pursuit

but coordinated with others, and thus operates according to laws that are mutually compatible.

In humans, of course, gaze is much more than a simple movement of the eye.[4] It serves to convey intent, to indicate shifts in attention, and to express emotion. The dynamic properties of gaze are especially interesting.

How to Simplify a Problem in Geometry

One of the problems evolution had to solve was how to move and coordinate the large number of muscles that appeared in terrestrial animals. This is called the problem of the "multiple degrees of freedom," a curious expression that takes into account, among other things, the fact that we have many ways to move a limb but at the same time constraints and, above all, a need to control many rotations. Hence the contention by the Russian physiologist Nikolai Bernstein that evolu-

---

(Opposite) Figure 5. Four elements among the repertoire of mechanisms of gaze control. (1) The first two, vestibulo-ocular and optokinetic reflexes, stabilize the image of the world on the retina; (2) saccades enable us to explore the visual world; (3) ocular pursuit maintains the image of an object on the fovea. These systems are interchangeable in case one of them fails. They work in cooperation but may also compete. This modularity is an apparent complexity that actually allows vision to play a major role in developing very complex and varied behaviors. The brain areas are in part the same as in figure 2. NOT is the nucleus of optic tract; EBN, excitatory burst neuron; DLPN and DMPN, dorsolateral and dorsomedial pontine nuclei; FEF, frontal eye field; SEF, supplementary eye field; CN, caudate nucleus; SN, substantia nigra; Thal, thalamus; PFC, prefrontal cortex. [Details can be found in A. Berthoz, *The Sense of Movement* (Cambridge, MA: Harvard University Press, 2000), and A. Berthoz, *Reason and Emotion: The Cognitive Neuroscience of Decision Making* (Oxford: Oxford University Press, 2006).]

tion has found ways to reduce the number of degrees of freedom. The complexity of our muscular apparatus — and its geometry — is indeed immense. Yet the ability to capture a prey depends on being able to move this machinery in a fraction of a second. Even before you grasp an object with your hand, you have to fix your attention, then your gaze, on it, which in turn calls on the six muscles that move each eye and the roughly thirty-two that move the head! Of course, we can also use peripheral vision to guide movement, especially familiar movement in a stable environment.

A good example of simplexity is the solution to the difficult problem of *noncommutativity of rotations*. Let us pose the problem in its generality before we see how it applies to eye movements. Take a die and rotate it three times in succession. Note its position after the final rotation. Then take the same die and, starting from the same initial position, subject it to the same rotations only in a different order. You will see that the die ends up in a different position. This difference is a result of the noncommutativity of the rotations. Now, the eye is a sphere that rotates constantly. If you stare at a point in space and make three successive ocular saccades to fix on different locations, then repeat the process, the noncommutativity of rotations dictates that the torsional orientation of the eye will be different in both cases. Yet the position of the eye in its orbit does not change, which can be verified by the fact that our perception of the world does not change. This remarkable property is due to a mechanism that compensates for the noncommutativity of rotations.

This simplex mechanism works in the following way: All the rotations of the eye have their axis in the frontal plane, which simplifies control considerably by reducing a three-dimensional problem to a two-dimensional one.[5] It is known as Listing's law. Johann Benedict Listing, born in 1808, a pupil and friend of German mathematician and scientist Carl Friedrich Gauss, was among the first to be interested in topology and to understand that it constituted a totally distinct field of mathematics. He had a talent for *le mot juste:* He introduced not only the word *topology*,[6] but also applied the term *complex* to aggregates of points, lines, and polygons.[7]

Which of our brain's mechanisms forms the basis for Listing's law is a subject that has given rise to much speculation. Is this property a result of the organization of the muscular apparatus or, rather, of central processing? It has been suggested that the central processing in the pathways that link the retina to the eye muscles explains why all the rotations of the eye take place around an axis in the frontal plane, which greatly simplifies gaze control. For example, the cortical structure most important for processing visual information about gaze displacement is the superior colliculus. It has been proposed that Listing's law is implemented in the neural networks of the superior colliculus by operations similar to what mathematicians call "quaternions." These operators are derived from calculations using "imaginary numbers," which would go a long way to simplifying the neural implementation of Listing's law.

However, detailed examination of the muscles recently revealed a very simplex solution: "pulleys" based on the following principle.[8] Each of the extraocular muscles is made up of bundles of muscular fibers. One of them provides the traction that triggers rotation of the eye in a plane. It crosses a ring that behaves like a pulley, modifying the eye's axis of rotation. This pulley is itself moved by another bundle of muscular fibers that (according to this theory) modifies the position of the pulley and thus the geometry of the eye's movement. If the theory proves correct, it would confirm a simplex mechanism that solves the problem of implementing Listing's law in an elegant fashion, though at the cost of some complication of the motor apparatus. This is what I call a simplex solution.

The Baby's Gaze

Even before an infant can walk, its gaze displacements enable it to explore the world and to exchange messages with near and dear ones. Organizing the repertoire of gaze movements is done by genetically programmed mechanisms and by epigenetic adjustments, which come into play during the baby's development, according to a relatively fixed schedule of maturation of brain function. An infant just a few months of age can already fix on a target with an average precision of 0.8 degrees—as opposed to 0.4 in an adult. This precision is thus superior to that attained by foveal maturation, which does not occur before the end of the first year. Seen this way, the eye movements are a simplex solution to the baby's motor deficiency. The ability to move the eye by jumps (saccades) to obtain a unique perception of the surrounding environment is by no means a simple process.

Chronologically, the fetus moves its eyes by week 16, and rapid movements begin toward 23 weeks. Ocular saccades as such do not appear until birth. They consist of two phases—rapid movement, then a little saccade of correction. In a child a few months of age, for some time ocular saccades are broken into several, "multiple" saccades (up to five or six). Although the mechanism is very sophisticated, it still takes a baby 15 weeks after it is born to be able to orient its gaze with a single saccade.

The latency of the ocular saccade is another parameter that varies with age: It diminishes over the course of years. The convergence is slower and the amplitude less in infants of four months, although their eyes can converge on targets situated at least 50 centimeters away. Control of vergence does not appear until four to six months and also corresponds to the development of accommodative convergence. Genuine ocular pursuit is established toward six to eight weeks. As for ocular pursuit, which makes it possible to keep a target moving on the fovea and even to anticipate its movement, it is not efficient before the age of one year, which corresponds to the development of the anterior parts of the brain (frontal and prefrontal cortex). The solution that consists of immobilizing the world on the retina by moving it allows the brain to economize complex operations for extracting the optic flow and is demanding in terms of processing. It requires a detour, but one that simplifies and optimizes efficiency.

Learning to See

Another original property of living organisms is the ability to learn complex sequences of movements. Learning to see entails grasping the order in which the environment must be investigated. Thus, when you work, you must see the tools you are going to use in a certain way. Generally, before you take hold of an object, you make an ocular saccade toward it. Looking anticipates and guides your action. In the same way, when you are driving, it is important to organize the targets of your gaze when you arrive at an intersection in a well-defined sequence. More generally, the ordering of action, like reasoning, is fundamental, and the brain uses many mechanisms of anticipation.

Learning activates the prefrontal cortex. When we have learned a complex series of gestures, the cortex is freed up and the movement becomes automatic. For example, learning to execute a series of saccades toward a target activates a zone of the cortex, the "prefrontal supplementary oculomotor area"; but when you execute familiar sequences, this activity disappears.[9] This finding is consistent with other observations made during learning of finger movements. How is that a simplification? Because a liberated prefrontal cortex is able to establish new links, make decisions, combine multiple data on the value of possible choices, arbitrate conflicts, inhibit behavior, perform logical operations, and participate in so-called executive functions. Accordingly, it is important that it be free and not taken up with learned or repetitive tasks.

The activity of the prefrontal cortex, which is involved in learning, is transferred to frontal structures — the supplementary motor area, or subcortical structures, the basal ganglia, and the cerebellum — which memorize the combinations required to make a gesture. The same is true for remembering trajectories and the "episodes" that mark them. The memory created by an event is formed in structures such as the hippocampus before it is transferred to other structures during rest or sleep: In this way, the hippocampus can form new memories.[10] We will return to this subject in chapter 10.

This redistribution of activity over the course of learning is a remarkable simplex solution. If every time that we wished to execute a learned gesture, we had to remobilize all the areas of the brain involved in learning, the brain would be continually occupied with the "programs" learned, and there would be no resources available for learning new movements.

Gaze: An Anchor for Action

In children as well as adults, gaze serves to anticipate grasping an object; it also helps to maintain balance and to guide locomotion. Moreover, there is also a strong relationship between development of the gaze-orientation response of the eyes or the head and a child's capacity to navigate in space using his entire body during locomotor activities. Ethologist Konrad Lorenz suggested this by affirming that the organization of a path requires a sequence of movements of orientation. Similarly, according to psychologist James J. Gibson, we represent space by combining successive movements of orientation. A child who cannot walk first imagines space in an egocentric frame of reference. When he is ready to take his first steps, he implements a remarkable strategy consisting of fixing (anchoring) his gaze to successive features of the room.[11] He then makes a movement by keeping his gaze fixed on this anchor, which for his memory corresponds to what today we call a local view. He then changes it and, progressively, acquires a series of local views of his motion. This succession of anchoring gaze movements makes it possible to build an intermediate representation. By using his gaze, the child will free himself and — probably thanks to a mental simulation of space — will be able to navigate without having to rely on gaze.

Don't Repair, Replace! Vicariance and Substitution

Evolution has constructed a repertoire of gaze movements: stabilization reflexes, saccades, and pursuit (see fig. 5). We have already stated that this specialization is one of the mechanisms of simplexity. It also has a remarkable advantage: In case of failure of one of the elements of the repertoire, it allows the brain to replace the deficient one with another element or a combination of elements. An example is what I have named the "saccadic substitution hypothesis."[12] When the vestibulo-ocular reflex fails, the brain uses the saccadic system and pursuit to create pseudo-reflexes (see fig. 6). It can replace a defective system with another element of the sensorimotor repertoire. This is one of the bases of the modern concept of remediation. *Rehabilitation* is repair of a defective function; *remediation* is the brain's way of creating a solution that replaces the failed system. For this, the brain uses redundant functions, or a combination of functions, of intact but unemployed circuits that replace the defective function. This is also the definition of *vicariance;* in early Christian churches, the vicar was one who replaced an absent curate. One of the advantages of simplexity is that it leaves open this possibility of substitution. A

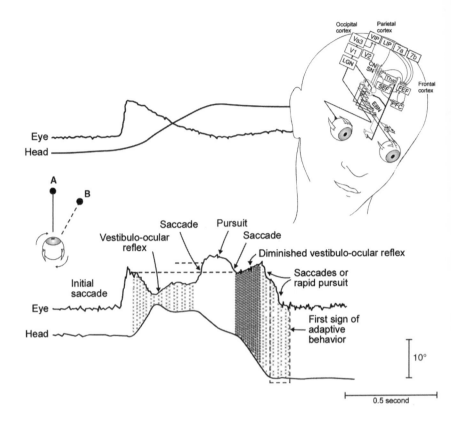

Figure 6. "Remediation" through functional substitution. The exceptional capacity of the brain to use the elements of the sensorimotor repertoire is illustrated here in the case of a sensory conflict. The upper part of the figure describes the horizontal movements of the eye and head when we gaze at target A then shift the direction of gaze by a combined eye and head movement to target B. The brain triggers a saccade, and the head moves. The vestibulo-ocular reflex stabilizes the image of the new target on the retina by automatically turning the eye in the orbit. The lower part of the figure shows what happens when Dove prisms—which invert the apparent direction of movement in the visual world—are placed in front of the eyes. In this case,

simple solution is fixed; it must proceed "without questioning," as the French army manual dictates. A simplex solution is flexible, adaptable, rich in possibility depending on the context, and often faster, despite the "detour" that it implies.

Who Is in the Mirror?

The exchange of gaze is a fundamental foundation of the complex problem of self-identity and social interactions. Its biological mechanisms include many aspects of simplexity. For instance, we take recognition of ourselves in the mirror to be the first sign of self-awareness.[13] When a child sees herself in the mirror, this recognition is (to my mind) based primarily on the look exchanged. The image in the mirror looks at the child, as the child looks at the image. Indeed, the child experiences a duality similar to what happens when we press our hands together and discover that each hand is both touched and touching. Here, the child is both seen and seeing. Self-awareness, which enables recognition of oneself in the mirror, is profoundly

---

(Figure 6 continued)

a saccade is triggered normally; but when the head moves, the vestibulo-ocular reflex is totally confused. For to look at the image of the new target on the fovea, the brain must produce a movement contrary to the normal reflex and do it twice as fast as the movement of the head. Here, the recordings on the figure show that, after several hesitations, the brain produces the required movement. It accomplishes this by inhibiting the reflex and replacing it by a combination of saccades and pursuits: A cognitive control replaces a nonfunctional automatic mechanism. This detour by means of alternative mechanisms for remediation, using vicariance, is an example of a simplex detour characteristic of the way living organisms have found solutions to complex problems. [Adapted from A. Berthoz and G. Melville Jones, *Adaptive Mechanisms in Gaze Control* (Amsterdam: Elsevier, 1985).]

linked to the infant's capacity to realize that the person looking back and the one looking are the same.

This ability to attribute a double function to seeing-seen also requires the (very young) child to be able to analyze the direction of the other's gaze.[14] In fact, recording evoked cortical potential (brain potentials resulting from the activity of many neurons following a visual stimulus) in four-month-old infants revealed a greater occipital (N170) evoked response than the presentation of images of faces whose gaze is direct. Moreover, even a five-day-old baby can tell the difference between a direct look and one that is averted. Brain imaging shows that direct visual contact activates the amygdala (an area deep in the lower part of the brain that is fundamental for emotion and attributing value to a stimulus), in addition to the fusiform gyrus (an area deep in the temporal lobe), which is involved in identifying faces. The baby's amygdala is already functioning, since it detects the gaze fixed on its own self. This could even be the neural basis of the "magnetic reaction" that arouses a reciprocal fascination between a mother and baby.

Exchanging looks is also evidence of the process of manipulating spatial frames of reference. Although the control of gaze movements is intrinsically egocentric, that is, governed by the perspective of the gazer, an exchange of looks requires both maintaining the point of view of the subject (egocentric) and shifting the point of view to the other (allocentric), which is the basis of empathy.[15] When I exchange looks with another person, I see him, but I am also in his shoes, looking at me. I call this remarkable operation "simultaneous multiperspective." A complete theory of gaze movements requires that we understand the role of the gaze of others on our own seeing. A solid theory of gaze interaction has yet to be formulated.[16]

My intuition tells me that this subtle and powerful game of intersubjectivity is also a fundamental property of simplexity. Through mimicry, contagion, and motor and emotional resonance, a child learns a lot. But despite the great theories of Sigmund Freud and of innumerable psychiatrists over the last 100 years, these exchanges contain something powerful whose relevance to what we learn about the world and ourselves we still do not comprehend. Especially as the construction of a self involves not only the past (autobiographical memory) and the present but also projection into the future.

Gaze and Empathy

Vision is a remarkably complex neuronal machinery, but in fact, its organization reduces the brain's complex processing. It is a simplex system not only because of the numerous mechanisms that simplify neural control, but also because it is a genuine tool for exploring the world and communicating with others. Even in nonhuman primate societies, interpretation of facial expressions and gaze orientation is important in identifying social position in a hierarchy or in identifying intentions. In humans, the ability to instantly recognize a face in a crowd is remarkable. It reveals the extraordinary efficiency of the neuronal system in detecting faces. The ability to examine a face and to recognize it may have developed very early, well before the appearance of *Homo erectus* or *Homo sapiens sapiens*, for the most evolved primates—deprived of language—would have relied even more than we do on facial recognition for social interactions.

Faces are reconstructed in the inferior portions of the temporal lobe, as evidenced by deficits known as prosopagnosia, induced in humans by lesions in the region that contains face cells, in the superior temporal sulcus; in the monkey this is in the inferior temporal lobe. Face cells respond to certain arrangements or combinations of specific features, like hair, mouth, and eyes. It has been proposed that these neurons be divided into four classes, based on the angle preferred by the cell (face, left profile, right profile, back). These neurons are sensitive to the gaze orientation of others, whereas other cells prefer certain kinds of faces.

All these neuronal responses derive from a broader function: social contact through gaze. The gaze of a social partner has a dramatic effect on the neurons of the superior temporal sulcus, which are sensitive both to the direction of the head and the orientation of gaze in the eye orbit. The neurons that respond to the gaze of the social partner have achieved a higher level of abstraction than the cells of classical faces. The temporal cortex around the superior temporal sulcus contains a field that is involved in social recognition.[17] The lesions of the median temporal lobe induce a social agnosia of gaze that is distinct from prosopagnosia and may be associated with difficulties in recognizing others' facial expressions.[18]

In the human species, exchanging looks is fundamental. This exchange is a crucial component of sympathy and empathy, which we have discussed in other publications. As we already mentioned above, eye contact immediately activates the amygdala, which assigns value to the objects in the environment and triggers flight or approach.[19] In one South American society, if a young girl looks at a young man, it means she has agreed to marry him. What a fine solution to a complex problem! We also know that a young child is fully capable of discerning the intentions of others based on where they are looking. This capacity—a fundamental mechanism known as "conjoined attention"—develops with age. We shall come back to this in the next chapter.

For example, say you show an infant a person sitting at a table that has two stuffed animals in front of it. The child is pretrained by watching the following scene: The person looks at one of the animals, then picks it up. The direction of gaze indicates the intention of grasping the animal. The scene is repeated until the infant stops paying attention to it, after which he is shown two scenes in succession: one where the person eyes the animal and grasps it (congruent situation), and another in which the person grabs an animal different from the one looked at (noncongruent situation). This last situation generally arouses the child's curiosity, which implies that infants really do use gaze direction to infer an intention.

This succinct analysis of gaze mechanisms suggests that simplifying principles are not situated uniquely at the motor or the sensory level. Rather, they bring into play high-order biological processes, such as attention, that enable selection and anticipation, which constitute two of the basic elements of simplexity.

4. Attention

Research on selectivity and capacity limits in perception was for many years . . . driven by the debate between early- and late-selection theories. These theories addressed fundamental questions about the relation between attention and perception. . . . While a few writers have contended that the questions were flawed to begin with, and others have proclaimed the unequivocal triumph of one theory or the other, many simply seem weary of the battle but uncertain of its resolution.

— *Harold E. Pashler*

Ticks care only about butyric acid and temperature. Their *Umwelt* is very limited! This is a primitive form of selective attention. In humans, attentional mechanisms are numerous and much subtler and also involve memory and context. But attention does not only depend on cognitive factors. Emotion, sexual desire, and motivations play very important roles in determining the focus of attention and its general properties. Finally, attention is inextricably linked to the problem of intersubjectivity, that is, our relationships with others. Already in animals attention can follow rules that are deeply embedded in social relations. One spectacular example is "imprinting," described by Konrad Lorenz, his collaborator Nikolaas Tinbergen, and others.[1] It is the name given to the fact that the attention of baby animals is captured by any mother-looking object during the small window of time called the "critical period."

For each species, the construction of very selective worlds is a fundamental simplification that aids in elaborating a simplex world we can control. Selection optimizes interactions between the living organism and the physical world. I assert that attention — a basic mechanism of cognitive processes in humans — is very closely linked to the general idea of *Umwelt:* We create world according to our needs. I suggest that over the course of evolution, mechanisms appeared that enabled us to adapt the repertoire of pertinent information based on the goals, desires, and beliefs of each individual, at every instant.

Our attention is characterized by the ability to alter our focus of interest. This switch may be provoked by external stimuli (exogenous attentional changes) or the intention of the subject (endogenous attentional changes). Moreover, the brain structures involved in these two sorts of changes are not the same. Attention, which is an expression of intentionality, is an original and elegant way of simplifying our interactions with the world by virtue not of simple mechanisms but of complex ones.

Among the principles that define simplexity, I mentioned above the detour via complex mechanisms. Attention raises all sorts of questions: Is attention serial or parallel?[2] Can preattentive mechanisms be dissociated from actual attentive processes? Is attention linked to gaze movements? Is it conscious or unconscious? Does it function as an attentional supervisor regulating decision making in the brain?[3] And, especially, is it ascending, that is, centripetal, consisting of a succession of processing steps from the sensory receptors up to the decision-making centers of the frontal and prefrontal brain? Or, conversely, is it descending, that is, driven by intentions, goals, and rules? Is it intermittent or continuous? Does it cooperate with working memory? Our objective here is not to propose a physiological treatise on the many mechanisms involved in attention, but rather to try and show that attention is a basic tool of simplexity.

"I Choose, Therefore I Am"

The philosophical context of the concept, and mechanisms, of attention is important to understand because of the fundamental role of attention in simplex processes for living organisms. For John Locke, George Berkeley, and their fellow empiricists, the brain receives sensory inputs that it combines and out of which it then extracts general properties. According to these philosophers, attention participates in this centripetal (that is, bottom-up), progressive selection schema, which goes from sensations to ideas. Attention is some kind of bottom-up filter mechanism, and the classical theory of the "unique channel" that takes into account the fact that we often can only do, or think about, one thing at a time seems to be a consolidating argument for this view of attention. This is partly true, but one must also consider the inverse projective (top-down) pathway, which leads from intention to the external world and, further, to the relationship established between the two.

However, the point of view of the empiricists was turned on its head by Edmund Husserl,[4] who maintains that we assign an identity to things a priori through *intentional* projective mechanisms. If we accept the idea that the brain interrogates and sees the world based on this prospective relation, then every intention corresponds to a new configuration of expected sensory inputs, and each act calls into play a different attentional activity. Here, attention is not only a filter but an expression of intention anchored in action. As we have seen, this anticipatory feature of brain function is one of the basic elements of simplexity.

One of the problems posed by the concept of attention is how to know whether its processes are conscious or unconscious. In the late 1800s, American psychologist and philosopher William James phrased the question of extent of the field of consciousness very elegantly. "Although we are besieged at every moment by impressions from our whole sensory surface," he wrote, "we notice so very small a part of them. The sum total of our impressions never enters into our *experience*, consciously so called, which runs through this sum total like a tiny rill through a broad flowery mead. Yet the physical impressions . . . affect our sense-organs just as energetically. Why they fail to pierce the mind is a mystery."[5] James pursued his train of thought, arguing that consciousness is interested in various elements of its contents to a greater or lesser degree, retaining some and rejecting others. For James, *to think is to choose*, which leads him to consider the Cartesian cogito and, so to speak, to replace "I think, therefore I am" by something like "I choose, or I select, therefore I am" (the formulation is mine).

One suspects that the problem of attention also entails that of free will. James suggests it himself by trying to dissect the major mechanisms underlying attention, namely, grouping of perceptions; opposition and competition; and finally, elimination. Contrary to the theories of the empiricists, James subsumes attention to the primary control of intention: "Attention, . . . out of all the sensations yielded, picks out certain ones as worthy of its notice and suppresses all the rest. We notice only those sensations which are signs to us of *things* which happen practically or aesthetically to interest us."[6] Here, what matters for us is that James intuited the existence of centrifugal mechanisms. A dual action is thus sketched: on the one hand, filtering and amplification in the afferent pathways; on the other hand, prespecification coming from the brain itself that projects its preperceptions onto the world.

If a single attentional mechanism controls our choices, we might properly speak of simplicity. The simplification that attention brings by focusing exploration of the world is subject to an apparent complexity by the fact that, as for all the major cognitive functions (memory, emotions), we are talking about more than one function. James knows it, and he proposes a classification of various forms of attention, prefiguring current knowledge regarding the existence of multiple attentional processes. He cites dispersed attention, sensory or intellectual attention, immediate attention, voluntary or spontaneous attention, and divided attention. Suggesting a combinatorics of attentional processes, he illustrates his idea by showing that lightly tapping a window arouses attention differently depending on whether one is awaiting a lover on a moonlit night or dreading the intrusion of a robber. For James, as for German physiologist and physicist Hermann von Helmholtz before him,[7] paying attention is a dynamic processes that requires a sustained effort: "The natural tendency of attention when left to itself is to wander to ever new things; and so soon as the interest of its object is over, so soon as nothing new is to be noticed there, it passes, in spite of our will, to something else. If we wish to keep it upon one and the same object, we must seek constantly to find out something new about the latter, especially if other powerful impressions are attracting us away." Finally, James turns to the physiological basis of attention. The first rule he proposes is the following: "For attention to be drawn to an object and to perceive it in its entirety, it cannot only be perceived by the senses; it must also be present in the imagination, that is, it must be represented twice."[8] At the time, this was a revolutionary proposition.[9]

Personally, I do not see attention as merely a mechanism grafted onto perception, like a general-in-command who decides what and whom to look at. It is a mechanism of anticipation that prepares for action, a mechanism that *configures the world for our actions and our intentions,* whose trace is found at all levels of the nervous system, from the most primitive to the most cognitive. It does not only concern conscious perception of real stimuli and "little perceptions"; it can also use the imagination. Recall that for Gottfried Leibniz, the idea of little perception refers to the fact that when, for example, we perceive the murmur of the sea, we are actually perceiving the sound of each drop of water, though in a confused way and without realizing it. Leibniz considers conscious perception to be a sum, the result of a process of integration carried out by a multitude of little perceptions all unbeknownst to us. For Leibniz, to some extent the soul has a perception — vague, at least — of everything that occurs in the universe. If we accept that most of mental life is unconscious, it is easy to understand why he is considered one of the major forerunners of the theory of the unconscious.

Let us take an example from research I carried out together with the group of psychologist Stephen Kosslyn.[10] If you stand before a disk that is spinning in a frontal plane, parallel to your chest, you will induce a vection, in other words, the illusion of bodily rotation: You will have the impression of turning in the sense opposite to the disk. This illusion provokes a modification of the vertical percept (or vertical subjective) that can be measured by positioning a little arrow in the center of the disk that measures the vertical percept. Now suppose that part of the disk is hidden, for example, the entire central part is black and immobile. If you try and imagine that the black part is moving in a sense contrary to the disk, you create a conflict between what you see and what you imagine. This mental imagery of movement modifies the vertical subjective, proving that what we imagine influences our perception of the vertical. The influence of attention on perception thus far surpasses the effect of focalization or amplification. Our brain can impose a priori interpretations on the world following its own intentions. You might say it makes rules, and each of us follows these rules based on our own experience, intentions, and culture. To me, this proactive function of the brain is a typically simplex mechanism.

Attention and Decision Making

The neural bases of different kinds of attention hold the key to the major mechanisms of simplexity. Readers pressed for time may wish to skip this section. But it is worth taking a closer look at the exceptional faculty of living organisms that are not prisoners of a fixed *Umwelt* but that have the possibility of choosing the world they would like to live in, using the many facets of their attentional system.

As mentioned above, there are several categories of attention: spatial, visual, sustained, selective, early, late, preattentive, divided, focalized, and so forth. Attentional effects have been found in practically all the visual pathways, and they also influence the motor systems, as predicted by the motor theory of attention. Amplification of the visual response with increased attention was discovered in the colliculus. In 1981, the neurophysiologist Vernon Mountcastle showed in the parietal cortex that the intensity of a stimulus response varies as a function of the level of attention. In the ventral system, the pathway from the visual cortex at the back of the brain to the temporal lobe, which analyzes the shape of objects, the environment, and living creatures, the receptor fields of the neurons of the inferotemporal cortex assemble around a visual target. The fact of directing attention toward a point in space increases the sensitivity of neurons of the parietal cortex to signals coming from the space that surrounds the target. The sensitivity to the tip of the target is increased in the neurons of the inferotemporal cortex.

Thus, attention induces very specific effects in the various parts of the brain that depend on processing governed by each part. Today, evoked potentials enable us to determine the temporal aspect of signal processing in the different cortical areas and to answer the question of whether processing is early or late. One example: A person is made to hear a sound in each ear and is asked to ignore what is going on in one ear. Recording of the cortical potential evoked by the sounds shows that the early waves (in particular wave N1) are modulated by attention. Primary variations in brain potential (P1 and N1) are greater when the stimulus is the object to which visual attention is directed. Other studies of evoked potential have emphasized the respective role of the magno- and parvocellular visual pathways in attentional processing of aspects of color and the shape of visual targets.

Here is another example of the high-order decisional and switching mechanisms involved in attention. Suppose now that you are presented with a cloud of dots, as when you turn the television to a channel without a signal. You would see a random cloud. If we were to assign each of the points on the screen a speed that, for a small group of the points, corresponds to the apparent speed it would have if it were on a sphere (a ball, for example), the experiment would show that the brain immediately perceives a sphere, whereas the dots are still on a flat screen.[11] You might then be asked to pay attention either to the movement of the sphere, which appears to be turning, or to the shape. Brain imaging makes it possible to see that networks of two very different areas of the brain are activated in these two forms of attention. Attention is more than a simple selection of sensory inputs. Analysis of the world differs depending on whether you are looking at two things as simple as the shape or movement of an object. The same thing happens when you focus on this or that feature of a visual scene: When you look at a face, different areas are activated if you are paying attention to the color of the face, its shape, or its position.[12]

Attention is therefore not just a selective filtering or amplifying mechanism. It is a genuine cognitive and motor function, rooted in action. It participates, rightfully, you might say, in decision making regarding action. Consequently, it makes sense to examine the relationships between the neuronal structures involved in attention and those where one assumes sensory motor decisions are made, as well as planning for action.

Another good example is the "decision to look." Let us do an experiment. Arrange three objects in front of you. We will call the object in the center C, and the objects on either side L (left) and R (right). Look at object C and make a saccade to one of the other two objects. To decide which object you are going to look at, your brain calls into play the centers of decision making in the frontal and prefrontal cortices.[13]

What type of mechanism might underlie this activity of the frontal cortex? The answer is provided by a task employed by neuro-physiologists.[14] It consists of presenting a monkey with a very simple scene composed of eight targets arranged in a circle, one of which is a different color from the others. In English, it is called *oddball*. The monkey is trained to make a saccade toward the target that is different (incongruent). The response is essentially one of detecting significant novelty in the everyday world. Recordings of the activity of the frontal cortex (the frontal oculomotor field) show that the response time to complete the task varies significantly. The neurons of the frontal oculomotor field have a special property: They are active for saccades in one particular direction, called the motor field, the spatial region that the neuron prefers. The activity of the frontal neurons slowly increases prior to the saccade. Depending on whether the monkey decides to make a saccade in the direction of the target or not, the neuronal activity will be profoundly different, reflecting the process of attention and decision making. It appears that the decision is made at the very moment the activity of the neurons involved reaches a threshold. If this theory is confirmed, it would imply an extremely simplex mechanism of decision making. The activity of these neurons would constitute a physiological marker of the precise instant the decision to look, and where, begins and ends.

In this context, looking requires deciding on a single target, that is, making a unique and drastic choice. Each saccade is a one-way decision. A mechanism in the brain known as *inhibition of return* prevents gaze from returning twice to the same target.[15] Some brain mechanism, still mysterious, keeps us from accumulating redundant information. Our perceptions are indeed well guided by the presence of objects in the world, but also by our intentions. Automatic mechanisms make the job easier for us by selecting what we see. More precisely, our perceptions are the result of two opposite processes, centripetal (bottom up) and centrifugal (top down). For reasons I find obscure, a proposal was recently made to name these processes "hierarchies" and "reverse hierarchies."[16] What is certain is that a brain mechanism avoids the accumulation of redundant information. The attention mechanism joins with working memory.

However, when we examine the world around us, at any given moment we pick up only a little information. This selection is accompanied by a fairly remarkable incapacity to detect certain changes. This phenomenon, which I have discussed in my book *Emotion and Reason*, is called *change blindness* and can be interpreted two ways: either as depriving us of important information about the world or, alternatively, as an economizing principle that assumes that the world is stable until proven otherwise. In any event, it is a simplifying principle that illustrates a conservative tendency.

Selective Ignorance

Many models have been proposed to explain the different forms of attention. For example, one possible process is selective filtering with suppression of nonpertinent information:[17] This is the inhibitory function of attention. Paying attention to one thing or event means ignoring others. I call this selective ignorance. But another interpretation suggests that information is not suppressed; rather, the selected region is freed from having to compete with other surrounding regions, which are not necessarily suppressed. This idea is interesting because it links the process of attention to more general processes having to do with discriminating objects in the visual world.

Another function attributed to attention, complementary to the first, is that of information filtering, or "selective attention," which chooses the signals most relevant to the action under way. This filtering, too, is the product of an astonishing property of the brain. Whereas the brain can process many data in parallel—this is one of the fundamental properties of vision—the frontal and prefrontal portions of the brain involved in mechanisms of decision making and judgment can only process small amounts of information simultaneously, and often only a single piece of information. This "single-channel" theory is a classic in psychology. It was shown long ago that the brain can only process one complex set of stimuli at a time. Today it finds application in studies of the process of economic and financial decision making in neuroeconomics.

Another form of attention is "divided attention," which restricts us to processing only a part of the available information. The mechanisms of divided attention are still not well understood, but they involve subtle processes in which inhibition certainly plays an important role.[18]

Similarly, a phenomenon called "attentional blink" clearly shows the existence of very robust mechanisms of temporal selection. When an experimental subject is asked to detect two targets (T1 and T2) in succession, a reduction in the ability to detect the second target, presented within a precise temporal window—a limited time interval—can be observed.[19] The percentage of correct detection of the second target with respect to correct detection of the first target depends on the delay between the targets.[20] When a subject is presented with this type of stimulus, a variation in positive potential (evoked potential) is produced with a delay — a latency — of 300 milliseconds. This P300 wave diminishes during the attentional blink as if attention were suppressed during this time. It is possible that this restriction in the number of decisions that can be made at any instant is a mechanism of simplicity, for it enables the brain to avoid being confounded by too much choice.

Attention and Vigilance

Attention is strongly linked to vigilance.[21] It is subjected to the action of major neuromodulatory systems that cause neuronal activity to vary based on learning, emotion, and context.[22] In the model of attentional circuits proposed by neurologist Marsel Mesulam,[23] several regions of the brain interact in the following way: The posterior parietal cortex establishes an internal "perceptual map" of the external world; the cingulate cortex regulates the distribution of motivational valence (the value placed on a reward); the frontal cortex (around the frontal oculomotor field, which controls gaze) coordinates the motor progress of visual exploration and grasping; and the reticular formation governs the level of alertness by means of the cholinergic, adrenergic, and dopaminergic neuromodulatory systems. These canonical areas are also connected to the orbitofrontal cortex and the inferotemporal cortex.

This entire network is endowed with the capacity of parallel processing and flexibility. Psychologist Michael Posner considers three networks to be involved. The first is the posterior attentional network (parietal cortex, superior colliculus, and a subcortical nucleus called pulvinar, which, among other functions, is also an important relay from the superior colliculus to the visual cortex), whose principal function is to orient attention toward a particular place in space. The parietal cortex disengages attention from a fixed target. The superior colliculus displaces the "attentional spotlight" toward the new target of interest. As for the pulvinar, it is also involved in defining the new target. The second network includes the anterior cingulate cortex and the supplementary motor area, which are active in numerous situations involving preparation for action and detecting events. The third network involves the noradrenergic system, originating in the locus coeruleus, a nucleus located in the deep-brain-stem areas. According to Posner, the posterior system is unconscious, and it is the intervention of the anterior system that jolts consciousness. The system of vigilance activates the posterior system and suppresses the anterior system, leading to a state of attentive alertness, but free of conscious content (called effacement of consciousness).[24]

In this model, as in that of Mesulam, attention results from a double process: activation of selected regions of the visual world and diminution of activity in neglected regions.[25] Based on this description, the simplification provided by attentional mechanisms has nothing simple about it. According to our definition, it is simplex because it obeys the principles we stated earlier: modularity, mutual exclusion, role of past experience, and so forth.

Biased Competition

In another model, called biased competition,[26] the neurons involved are the object of competitive processes. For example, there is competition between the centripetal effects (originating in the visual centers), which progressively enlarge the visual receptor fields of the neurons, and the centrifugal effects (originating in the frontal and prefrontal areas), which shrink them. This model is not that of the attentional spotlight sweeping the visual field like a projector light in a prison yard. It is characterized by the following constraints and properties:

- Only a limited number of objects can be represented simultaneously, considering our modest capacity for processing different pieces of information at the same time (single channel). This puts us in a typically empiricist position, since this limited capacity results from the existence of a bottleneck and not of a unified point of view.
- Attention cannot be reduced to ascending, centripetal, or progressive selection mechanisms. It is also guided by descending and centrifugal impulses. I often use the term *projective* to designate this property of the brain as emulator. Attention directly influences (and in a centrifugal way) processing of attributes of visual forms at the earliest levels of this processing.[27] Others give priority to the simplifying character of a special mode of encoding — indexing (see the theory of cognitive scientist Zenon Pylyshyn[28]). The competition between centrifugal effects and centripetal effects is integrated in several brain areas, and a winner emerges.
- Competition derives from neuronal processes of inhibition. One of the problems posed by the theory of biased competition is that of the link between all the representations of objects and of the world, since the different populations of neurons involved in competition must know that the object is unique. I mentioned above how we construct a unified and permanent perception based on a world broken into diverse components and attributes. This problem, also called the "binding" problem, applies, obviously, to attention. One attempt to resolve this problem is the theory of feature binding (color, shape, and movement) formulated by the psychologist Anne Treisman.[29] Features are analyzed in parallel by low-level, preattentional processes, in every visual scene. I am referring to the attentional displacements imposed by the subject and his intentions that control the binding

of different bits of information and help to construct a coherent representation. These features are integrated with reference to a neural map of the locations that encode the position of the elements in the world and maintain the links with the neuronal map of the attributes. The parietal cortex is one of the possible preferred locations of this integration.[30] Thus, during visual exploration, a mechanism of attention based on features—color and shape, for example—prepares the terrain for focal attention, which itself ensures orientation toward the object of interest.

Processing visual information is not done only in a serial and hierarchical way. It relies on parallel involvement of numerous feedback or centrifugal control mechanisms and mobilizing processes of selection at each junction.[31] Modern neuroscience continually sheds new light on mechanisms of attention, and by the time this book is published, new theories will have been proposed. However, our goal here is not to review the most recent discoveries regarding this or other mechanisms but rather to present a few that help us to explain the concept of simplexity in living organisms. In any case, attention modulates activity at the first stages of visual processing. It is not content with amplifying visual inputs; it profoundly modifies the inputs themselves. Depending on context, it acts on the spatial processing of objects, their attributes, or a combination of the two. The fundamental aim of attentional processes is to reduce the complexity of analysis of the world and its relationships with the acting body—a genuine mental posture in the sense of preparing for action. The variety of attentional mechanisms reflects the work achieved by evolution to simplify neurocomputation and to spare the brain having to process overly complex problems. It is characteristic of simplexity.

Attention and Emotion

Emotion is a powerful guide for attention. We do not only have a cognitive brain that analyzes spatial properties of movement, shape, direction, and content from the world around us and applies preperceptions to it. We also have a limbic brain that attributes value to things and to people. Attention is thus not simply guided by sensory properties. It is also strongly guided by reward and punishment, danger and pleasure, and value, which can be concrete or imaginary, linked to a memory or a bet, an assessment of probability, like winning at lotto. We have already mentioned that one area of the brain plays a particularly fundamental role in the rapid analysis of the value of shapes and objects that we encounter: the amygdala. Here we will expand the description of its properties. In less than 100 milliseconds, this complex neuronal structure assigns a value to a perceived shape. For example, if you are walking in the forest and you spy a wiggling object, the amygdala very quickly gives this object the status of snake, that is, a dangerous object. This perception then triggers a fear or flight response. Only later, after 200 or 300 milliseconds minimum, when the image of the snake has been analyzed by the temporal cortex and a finer-grained apprehension of the shape of the head, for example, has taken place, are we able to say that it is not a viper but a harmless garter snake.

The amygdala senses what is important. Among other functions, the central nucleus of the amygdala is involved in learning orientation toward a stimulus that is biologically significant. It influences the structures (for example, the striatum) that control behavior. What is at issue here is an influence on the mechanisms of orientation guided by a goal (intention) and not just orientation guided by a stimulus. The pathways involved include the following: the amygdala, the substantia nigra (a dark area in the middle brain), and innervation of the striatum by dopamine, which modulates the orientation response.

Lesions of the circuit that links the amygdala, the basal nucleus, and a deep-brain area called the substantia innominata govern divided attention, which also involves the medial prefrontal cortex. Here again, evolution has found an efficient solution that enables us to survive in our *Umwelt*. Nor is it a simplistic solution, for the amygdala is anything but a locus of simple operations.[32] It is a simplex solution to a complex problem: to identify, among the vast variety of shapes that surround us, those that risk putting our existence in danger and to trigger an appropriate response.

Development of Attentional Processes

Attention, or rather attentional processes, appear gradually in the infant over the course of ontogenesis. They are associated with cognitive development that makes it possible to choose from the world and, no doubt also, from memory, relevant information for each person at each instant.

We know that the attentional processes that enable rapid identification of others or imitation are precocious. These mechanisms, which do not require very complex processes, are implemented very early in babies. Naturally, the debate on the developmental calendar of infants is not entirely resolved. Accordingly, child psychologist Pierre Mounoud turns Piaget's proposition on its head and affirms that development proceeds from thought to action:[33] Babies come into the world with a ready-made, initial representational system, which explains their exceptional abilities. The ability of a four-year-old child is defined by saying that the infant activates or inhibits representations at will — whence the importance of inhibition and attentional mechanisms in the development of these competences and of logical thought.[34]

The frontal and prefrontal cortices, which are essential for selection, decision making, guiding action, and executive functions in general, develop later.[35] However, the questions related to ontogenesis do not only concern the calendar. There are many competing theories regarding the mechanisms that come into play during maturation of cognitive function. Thus, for some, a cognitive capacity such as attention results in maturation of a single region, for example, the dorsolateral frontal cortex. For others, maturation is the product of a change in the interaction of several regions, for example, between the dorsolateral frontal cortex, the parietal cortex, and the cerebellum, which enables us to identify and to find objects. Finally, it is possible to imagine that the essential mechanism consists in learning: The addition of the inferior parietal cortex to a network endows it with new capabilities vis-à-vis visuomotor tasks. In my opinion, these three theories are not mutually exclusive. However, the development of attention is not necessary even for very complex behavior like imitation: A very young baby can mimic the facial expressions of its parents, even though it cannot see its own face, which would seem to suggest the existence of innate mechanism of intermodal encoding and transfer between what is perceived and what is produced.[36] The baby can also immediately imitate emotion, called emotional contagion. These are simplex mechanisms, as we shall see below, of the ability to actively simulate the actions of others by involving one's own body and not only an abstract processing of information. This "embodiment" of perception is a fundamental mechanism of simplexity. I will only take one example from all the complex history of the development of attention. The appearance, toward 12–14 months, of "conjoined attention," mentioned above, represents a fundamental step. The infant, already capable of directing its attention, now seeks to share it with others. He will not only point at targets of interest but "show" or "designate" these targets to others. (The neurologist Jean-Denis Degos has found a specific impairment of this capacity in patients with parietal lesions.)[37] Doing this arms the child with one of the basics of social life. Currently, no robot can achieve conjoined attention with another robot. The early character of the maturation of attentional functions can be measured by dynamic analysis, which has revealed that the neural processes of perceptual integration that are present in the gamma band (40 Hz) in adults are also present in infants.

The ability to construct an illusion, that is, to imagine the world and to project interpretations onto it—one of the properties of simplexity—develops very early in the infant. A very promising avenue in this area is the analysis of the genetic basis of attention. There are, in fact, major differences in performance and cognitive strategies between groups with different genotypes during an attentional task, as well as between the sexes. We expect this area of human genetics will prove fruitful in the years to come.

5. The Brain as Emulator and Creator of Worlds

The world is infinitely complicated, yet the only way to investigate it is to ask questions. . . . The questions imply the answers and act as a "format" for converting simple structures into meaningful information. In this way, intention and meaning are imposed by the observer. To be a good observer requires being able to ask relevant questions of nature or, to put it differently, to have an interface that makes the world appear sufficiently simple to ensure survival. We need simple and effective geometries for interfaces, in contrast to the theories inevitably offered by physics.

—*J. Koenderink*

Our analysis of simplexity is guided by the central idea that not only action but also the *act* (a much broader concept, which I have discussed in several previous books) must be at the center of all analysis of the functioning of living organisms. To act, the perceiving brain relies on simplifying principles. Psychology and the neurosciences have now established profound links between perception and action. For this reason, I have introduced the term *perc-action* in my courses at the Collège de France. The simplifying principles, too, link perception to action. Of course, it is impossible to describe all the simplifying principles that evolution has devised: We discover more every day.[1]

Phenomenal Perception

Suppose that an image of two balls is projected onto a screen: a first one moving, the other still. The moving ball comes into contact with the other, which immediately starts moving in the same direction as the first ball. We have the illusory perception that the movement of the second ball is caused by the first one, which of course is not true. The brain has constructed a percept of physical causality. The Belgian psychologist Albert Michotte — often copied and rarely cited — conducted numerous experiments to test out his theories on what he called "phenomenal causality."[2] For Michotte, this impression of physical causality is a phenomenal given, which exists sui generis. It requires that two movements of distinct objects be integrated into a total, temporally contiguous unity and ordered in a way that the first is dominant and determines the point of reference of the second.[3] Here Michotte concurs with those for whom meaning is contained within perception itself: "The expressions which are used to describe this experience, far from investing it with a meaning or constituting an 'interpretation' of it, are indeed simply a translation into conceptual terms of what, at the phenomenal level, *actually does occur* [italics author's]."[4] A cognitive or logical interpretation is unnecessary to apprehend the global properties of the causal dynamics of relationships between objects.

Another psychologist, Roger Shepard, provided a totally original interpretation of the properties of apparent motion. The simplest case of apparent motion is that in which light is projected alternately left and right onto a screen. Under this circumstance, one perceives motion as if a single light were crossing the screen. The perception of the natural trajectories of animals — birds, for example — is also organized not as a function of particular characteristics of this or that animal, whose possible trajectories are many, but based on simple kinematic laws, in other words, laws that take into account the velocity, acceleration, and geometry of trajectories, in contrast to dynamic laws that concern, for instance, the forces at work. Perception takes in the continued existence (permanence) of the object and hypothesizes rigid movement — the simplest possible — which joins the first view to the second in a way that is compatible with the point of view of the observer or the perceived object.

Dreaming and Newton's Laws

The brain can predict the behavior of a moving object in the world and eventually catch it by virtue of "internal models" of the laws of physics. This proposition has sometimes been called "naïve physics." We have now been able to show that humans do possess neural mechanisms that allow the brain to simulate and use Newton's laws, for example. The reader can replicate the experiment. Pick up a bottle of water or a heavy object in your left hand and let it fall into your right hand. You will notice that your right hand does not lower when it encounters the weight. This is only possible because the arm muscle, the biceps, has produced a force in anticipation of the contact of the weight that is exactly what is needed to compensate the force of the impact. We postulated that the brain can only do this if it has an internal model of the acceleration of the object and Newtonian mechanics. We went to space because in a space station a ball will fall at constant velocity. We observed that even if vision detected that the ball was falling at constant velocity, the brain still produced the anticipation. We conclude that these programs are deeply inscribed in the brain's neural networks, and it takes some time for it to adapt to a new condition.[5]

Today, we know that many of the laws of physical behavior of objects in the world are internalized, that is, that the neurons in the brain have properties that enable it to simulate these laws. That is why, when you dream, you have the impression that a suitcase is heavy, that you are having trouble climbing a hill, or that you are forcing open a door — all in the absence of any information about the external world. In other words, when dreaming, you do not need the real world to simulate all its properties.

The brain can also anticipate the behavior of moving objects and people. Imagine that you are looking at a series of photographs of a person who is throwing a ball or jumping over a chair. And imagine that you are interrupted at photo N and that you are asked to choose among photos showing the most plausible next step in the movement (photo N+1). You will have to predict the trajectory of the ball or the jumper. Findings from this kind of experiment show that prediction is made very accurately based on the laws of Newtonian physics, thanks to internal models of the laws of mechanics and of the properties of the body.[6] Similarly, the brain very rapidly detects impossible movements on film, that is, ones that do not follow the laws of natural movement. It can also quickly distinguish a biological movement, even if the only information given is a few points corresponding to the person making the movement.

Shortcuts

According to Shepard, when there is little time to react, the perceptual system takes shortcuts: It does not, for example, employ analytical methods that respect Euclidean space, geometrical kinematics, or nineteenth-century French mathematician Michel Chasles's theorem of rigid-body displacement. This means, roughly speaking, that when we wish to go through, or imagine, the path between two solid objects on a three-dimensional surface, and if this path exists but we have very little time to do it, we will chose a geodesic path, that is, the shortest curve on the surface in question (like airplanes that fly from Paris to New York). Even if the short path is impossible, we can break the rules of rigidity and, if possible, modify the form of the object. When speed is of the essence, perception can break the rules that usually guide its functioning and create what I call shortcuts.

More generally, the complexity of the world can easily be sidestepped if the brain constructs a certain coherence between components of reality. In a commentary on biologist Jakob Johann von Uexküll, child psychologist Henri Piéron indicated that, paradoxically, the natural environment is easier to analyze than certain artificial environments (for example, a cloud of points) because it contains shapes (gestalts) that have little meaning but are instantly recognizable.[7] These shapes are recognized by "analyzers" in the brain that are specialized in recognizing animal forms (fusiform gyrus), body parts (extrastriate body area), environmental scenes and buildings (parahippocampus), and natural movement. I extend the term *shortcuts* to include these solutions found by the brain to interpret the world. Psychologist James J. Gibson had the same fundamental intuition when he wrote that the brain detects "affordances."[8] According to Gibson, the brain does not measure all the physical parameters of objects (luminance, contrast, dimension, and so forth), but rather the relationship between the perceived object (an armchair, a door) and an action in progress (sitting down, crossing the threshold). This idea was also proposed by the French physician Pierre Janet in the early 1900s.

The Brain as Emulator of Reality

The brain thus makes hypotheses. But it does not just simulate reality. It emulates possible worlds. For example, the brain assumes that objects are rigid, and it transforms the perceived world to make it as symmetrical as possible, at the cost of (perceptual) deformations of physical reality. The phenomenological brain possesses its own laws of interpretation. Shepard proposed that *perception is a hallucination directed from without:* Mental imagery and certain forms of thought were just perceptions simulated internally, at a very abstract level.[9]

Jan Koenderink, a Dutch physicist devoted to the study of human perception, who was among the first to suggest that the brain does not process visual or tactile information according to Euclidean geometry, believes that a normal brain produces hallucinations: Perception is not a representation of a world that exists independent of us. Koenderink writes: "Perceptions are not in the brain, nor in the world, they are in experience."[10] Moreover, they differ among individuals. "Variations among observers are huge. I find differences in depth range up to a factor of 5. When observers discuss the pictorial objects they tend to agree; the often enormous individual differences typically go unnoticed." He even finds differences in perception between individuals that reach as high as a factor of 70 and writes: "This is an amazing range that again typically goes unnoticed when observers discuss what is in the picture." These differences are linked to the fact that, at every moment, we seek to simplify by selecting what is relevant for us in the perceived world; we also emulate perceived worlds.

These ideas are not new. The Austrian scientific genius Ernst Mach began his 1893 treatise on mechanics in the following way: "The most direct, and in a sense the most important, problem which our conscious knowledge of nature should enable us to solve is the anticipation of future events, so that we may arrange our present affairs in accordance with such anticipation . . . and thus to draw inferences as to the future. . . . We form for ourselves images or symbols of external objects; and the form which we give them is such that the necessary consequents of the images in thought are always the images of the necessary consequents in nature of the things pictured. In order that this requirement may be satisfied, there must be a certain conformity between nature and our thought."[11] The brain really is an "emulator of reality." A fundamental aspect of simplexity is this creative activity. It solves the problem of the complexity of the external world by producing perceptions compatible with its future intentions, its memory of the past, and the laws it has internalized about the external world. It creates a veritable *Umwelt*. But it also makes mistakes.

"Déjà-Vu, Déjà-Vécu"

As already noted, a basic property of simplexity is anticipation, which is in part carried out by top-down perceptive or motor (centrifugal) influences. I have mentioned the hypotheses of rigidity and symmetry that we make, which determine the properties of what we perceive, either through sight or our other senses. Moreover, when we perceive an object under blurry or badly lit conditions, we attribute to the object properties borrowed from the library of objects stored in our memory. This is how sometimes, at dusk, you might mistake a rock for an elephant. The neural basis of this categorization, which gave rise to the theory of object model verification, is now known. The role of the prefrontal cortex in the phenomenon has been demonstrated,[12] but memories are stored in several other areas of the brain. American-born Canadian neurosurgeon Wilder Penfield identified areas in the temporal cortex that, when stimulated electrically in epileptic patients, evoked impressions of déjà-vu/déjà-vécu (*déjà vu* refers to an imagistic recollection of a scene, whereas *déjà-vécu* adds the impression of having fully experienced a life episode). This corresponds to so-called "episodic memory," which is the memory of a life event that includes sensory aspects, action, and spatial information about where it happened. Episodic memory plays a crucial role in autobiographical memory. Patients would say, "I feel I am sitting in my mother's living room." It is not only objects that are evoked by memory but entire lived scenes. This suggests that the projective brain is capable of evoking a complete scene to interpret the world as it is perceived and lived at a given moment in the past. Memory imposes models of interpretation. It is also a source of familiarity.

Resonance

The concept of resonance is a good illustration of the predictive character of perception. Gibson proposed that perception is a sort of resonance between the expectations of the nervous system and the invariants that it extracts from the external environment. This idea, also put forth by Karl Duncker in the 1930s, was confirmed by Shepard.[13] Rather than maintaining, like Gibson, that an organism recognizes "affordances" that are latent in sensory space, Shepard proposed that an organism is perpetually attuned to sensory configurations that are meaningful for it and closely linked to its environment. This is a striking premonition of the mirror neuron system, discovered by Italian neurophysiologist Giacomo Rizzolatti.[14] Resonant systems have three types of properties: First, they respond in different ways to the same stimuli, depending on their tuning; second, they are excited in different ways; and finally, they can be made to resonate by transmitting a simple impulse within their own structure. They can also resonate when sent a signal different from the one to which they are tuned if the signal has a certain relation — for example, harmonic — to the tuning frequency.

This capacity to make internal systems resonate through stimuli close to natural stimuli explains certain properties of "perceptual filling," that is, the brain's capacity to continue to see the external world even in the absence of signals. This is frequently the case with vision, in the ability to read a word even after a letter has been removed from it, or with faces. The idea can be generalized by saying that it is known that the nervous system resonates with very particular sensory elements such as sounds, light, or the displacement of a simple line; but the nervous system can also resonate with higher conceptual categories, such as shapes and faces. *Resonance is a simplex property* that makes neurocomputation much more economical.

Redundancy

Redundancy occupies a special place in the list of simplexity principles. At first blush, particularly for proponents of signal theory, like information-technology pioneer Claude Shannon, redundancy appears to be of little utility because it clogs channels. Yet redundancy has been used in transmission to avoid errors in mitigating the effects of noise. In biology, redundancy plays a role in sensory processing. Neuroscientist Horace Barlow notes: "My leading idea was the same as [gestalt theorist Fred] Attneave's: as he put it, 'the human brain could not possibly utilize all the information provided by states of stimulation that were not redundant.'"[15] In particular, redundancy aids prediction and anticipation: "Knowledge of the redundancy of the messages from the environment enables such a pattern to be identified in its early stages, and . . . makes prediction possible."[16] Redundancy is also important for inductive and inferential reasoning.

This property of redundancy is not limited to perception. It is fundamental to living organisms at all levels. An example of the biological implementation of redundancy is found in bacteria, where two pathways have been found to correspond to the two principles of diversity and redundancy. Biologist Daniel Koshland has observed: "Mother Nature follows the principle of redundancy by selecting a simple mechanism or module as a building block for a complex system and then using that module over and over again in other systems. . . .

Just as very different chemicals can be created by different permutations of protons, neutrons, and electrons, so can different biological species be constructed from similar receptors, enzyme pathways, and membranes."[17] This "two-component pathway" exists in both bacteria (prokaryotes) and plants (eukaryotes, living systems whose cells contain a nucleus). Consequently, it probably also exists in humans.

Around 1960, the great neurophysiologist Vernon Mountcastle proposed that, in the cerebral cortex, functional architecture is repeated many times, but it is the connectivity with other areas that defines the functional specificity of each area.

The Critical Period, or the Weeping Camel

One of evolution's most extraordinary solutions for simplifying brain function is undoubtedly the notion of "critical period." These periods consist of precise moments in development. Indeed, in baby animals, as in human infants, numerous sensory capabilities become established in an episodic fashion. For example, one critical period—which occurs in the first weeks after birth—concerns the development of the cellular properties of the visual cortex. A kitten blinded during this period will have a deficient visual system. Similarly, a kitten transported passively during this critical period will see poorly, for activity is fundamental in establishing sensorimotor and cognitive function.

The most famous example is "imprinting," mentioned in the last chapter, for it fixes in memory a form that thereafter is synonymous with mother.[18] Imprinting is so powerful that goslings will follow any object presented at the crucial moment. What is at work here is not simply a specialization localized to one sense but a complex process that enables the animal to identify shapes in the external world that are meaningful—its *Umwelt*, so to speak. My idea is that this emergence of functions over time simplifies relationships to the external world because it saves the brain having to establish different functional networks in parallel. The processes that underlie the critical period are complex and even occur at the molecular level, but they participate in simplexity.

Recently, it was suggested that there is a critical period for locomotion that establishes complex coordination between the production of locomotor rhythms and mechanisms of postural control.[19] I suggest that there are also critical periods in the development of cognitive function, such as the ability to change point of view, which according to Piaget takes place around the age of 7 or 8. This age represents a critical cognitive period.[20] I have proposed that if a child is taken between 7 or 8 and 12 years of age, say, secluded, and taught only to believe in one kind of religion or is educated in the hatred of others (as has been and is still done in sects or in communities that train terrorists), the child will be marked for life and grow to have a fanatical view of social interaction instead of a tolerant attitude toward others. Generally speaking, development is organized in very precise temporal sequences, windows that open and close. The advantage of this organization is that it simplifies the establishment of sensorimotor and cognitive function.

Let us conclude this brief evocation of critical periods with the charming story of the weeping camel.[21] A family of nomads lived self-sufficiently in Mongolia, where they raised the camels that are a part of their daily life. One camel had just given birth, but with great difficulty. As a consequence, the mother camel showed little interest in her baby and refused to let it nurse. Tradition holds that the playing of a violin can sway a camel and reconcile her with her calf. So the nomad father sent his son to a neighboring town to look for an appropriately specialized musician. The musician came to the village and played to the mother and baby camels. After a while, the mother camel began to weep and gradually sidled over to her baby, in the end allowing it to suckle. This moving story is strange but true.

How are we to understand it? Owing to the difficult birth, the link (imprinting) between mother and infant could not be established at the critical moment that enables bonding. That is why the mother denied her calf milk. However, it seems that the tonalities and rhythm of the music contain signs — or some musical form — that enable the crucial link to be produced anew. Listening to the music and seeing the baby camel nearby, the mother camel associated the sound of the violin with her baby, and under the influence of the music accepted her infant.

On the whole, the neural basis of the critical period is still mysterious. Yet we understand quite a bit about vision, including some of the molecular mechanisms behind it.[22] In the chicken, for example, a special structure of the anterior brain — the intermediate medial hyperstriatum ventrale — is the seat of transformation of postsynaptic density (the number of receptors located at the level of the synapse between two neurons on the side of the receiving neuron), which may be responsible for memorizing visual stimuli.[23] A recently discovered mechanism involves inhibition of the neurons of the primary visual cortex. I note here that this is yet another mechanism of selection by inhibition that modulates the excitability of the neurons of the visual cortex during the critical periods.

A second recently discovered mechanism, the homeoprotein OTX2, influences the maturation of neurons involved in plasticity in the mouse after closing and opening of the eyes.[24] This work, along with other efforts, is the first proof that transmission involving GABA (gamma-aminobutyric acid) is necessary for plasticity in vivo. The findings confirm that the temporal progression of the critical period can be controlled by inhibitor interneurons. However subtle these mechanisms are, the remarkable fact relevant for simplexity is that the critical periods for sensory, motor, and probably also cognitive function occur according to a time schedule that permits their precise coordination. Little by little, the brain is thus able to construct the global integration of behavior. The temporal organization of functional development in life is one of the fundamental features of simplexity.

Identifying Objects

One of the most fascinating questions in physiology is that of the unity of perception, that is, the fact that we perceive objects or animals as entities. As early as the Middle Ages, objects were seen as possessing distinct characteristics referred to as *essentia* and *accidentia*. Without going too deeply into this field of research, let us briefly survey what might be called the debate over the unity of perception. Certain psychologists, such as Gibson,[25] Michael Turvey, and Michotte, maintain to a greater or lesser degree (and from varying points of view) that we have direct access to the properties of objects in the environment. This idea appears to have been called into question by the fact that sensory detectors, in all animals and in humans, deconstruct reality into very simple elements: components. In this way, the neurons of the visual pathways encode shape, color, and contrast in a very specialized manner. As I have stressed in this book, we need a new theory that links the fragmentation of perception in the sensory system with its global and predictive character. The question is a difficult one, and as yet we have not answered it.

A common feature of all the theories and discoveries on this subject is that the brain has access to several mechanisms for simplifying identification. Today, we assume that synchronization of oscillating waves in the "gamma" frequency band (between 40 and 150 oscillations per second) and the temporal coincidence of activities in the different centers that encode the relevant attributes may be important for perceptual linkage, that is, for reconstructing the unity of a deconstructed object. Already at the level of the visual cortex, there are mechanisms that permit early identification of some properties of objects. Spontaneous patterns are created that prepare analysis of the visual world.[26] Shape, color, luminance, and movement are themselves special groupings (spatiotemporal correlations) of signals, and not primary components. There is a continual tendency to form ever more complex arrangements (arrangements of arrangements).[27] For example, there are more configurations in a black-and-white image measuring 16 by 16 pixels than there are particles in the universe! We have not yet figured out how the brain assigns unity to worldly objects (or to our own body, which we perceive as an object). Solving this problem requires that we understand the purpose of our senses and the simplex mechanisms that allow them to apprehend the world.

6. Simplexity in Perception

This chapter presents a few examples of simplexity in perceptual systems. The purpose is not to be exhaustive but rather to illustrate the various facets of simplexity. The reader will find here and there some of the general principles that have been put forward in previous chapters.

Encoding the Continuous by the Discrete

A remarkable property of the nervous system is its ability to encode continuous phenomena by pulses, that is, discrete signals. For example, the length of a muscle is encoded by a pulse frequency fired by receptors known as neuromuscular "spindles." When the length of a muscle varies, these delicate receptors—in series with the muscle motor fibers—fire pulses whose frequency is proportional to length and velocity of the change in length. Rotation of the head is encoded by variations in the firing frequency of sensory cells in the vestibular organs of the inner ear that detect acceleration. Frequency coding of pulses is a general rule of many sensory systems. But the patterning of coding of pulses (their time distribution) is also a common mode. Finally, the coincidence of discrete pulses and synchronization by oscillations are also variations in the encoding of the continuous by the discrete. The advantage of the *detour* carried out by discrete encoding is to transform an incredibly complex process into a simple response, common to all sensors. This common code, this shared language, "simplifies" the work of the brain and at the same time allows a remarkable increase in the various modes of processing of information. We see this same principle at work in the nonliving world in telecommunications and, more generally, in the proliferation of digital representations of reality.

In addition, in many cases, this transfer to discrete encoding is accompanied by spatialization. For example, the acoustic sensor in the ear (the cochlea) encodes aerial vibrations produced by sounds in the form of a vibrating membrane (the basilar membrane). This membrane behaves as a spatial frequency filter because, along its axis, successive portions of the basilar membrane vibrate with different frequencies. Imagine singers aligned in a row producing sounds of increasing sharpness. From one end to the other of the cochlear spiral, sensors detect the movements of the basilar membrane and encode increasing frequencies. The use of space to classify the frequencies along a membrane is a very elegant solution that avoids having to do complex neural calculations.

Controlling Uncertainty and Randomness

The world is full of uncertainty; so are brain processes. Randomness and stochastic (that is, probabilistic) variations occur at all levels of functioning of the brain, from the synapses up to multisensory integration and even the cognitive processes of decision making. But probability has simplex qualities, for the apparent confusion that it introduces masks great advantages.

As we have seen, the idea of discrete encoding involves probability. When the ionic equilibrium of each side of the neuronal envelope (the cell membrane) reaches a certain threshold value, neurons emit action potentials. This in turn produces a small storm of ions that release potential along the axon to inform neurons farther along the event. This neuronal discharge, which is the basis of discrete encoding, is probabilistic. That is, it responds largely to the laws of chance. Recordings of neuronal activity show random-type fluctuations that appear to introduce noise. This noise has, for example, the interesting characteristic of sharpening the sensory signal and improving the reliability of detection by taking into account past information and predicting the future state of the sensor. It is a sort of Bayesian process (see chapter 2).[1]

More generally, the synapse — the link between the axon of one neuron and the cellular body (soma) of the next neuron — is where processes of immense complexity take place. These processes are no doubt simplified by laws (which we do not yet understand) that ensure transmission of signals between neurons not by electrical waves but by chemical interactions involving receptors. For a long time it was believed that all this machinery was relatively stable. In fact, the release of neuromediators obeys laws of probability, and the movements of receptors within the neuronal plasmic membrane reflect Brownian motion, that is, the receptors are in constant agitation. Thanks to this random jiggling, the receptors can also move and explore regions of several square microns in periods of a few minutes. Such phases of rapid movement alternate with periods during which the receptors are confined — quasi-immobilized — and diffusion is reduced by several orders of magnitude. The stability of the synapse and its physiological interactions thus does not result in molecular stability but rather in dynamic stability of complex molecular assemblies.[2] It is the result of a dynamic (statistical) equilibrium, where the molecules are in continual movement in "microdomains." Remarkably, these movements are regulated by precise physiological processes that use chance to modify the state of the synapse depending on the context. In this way, the probabilistic nature of the process contributes to the flexibility of the device. As this example shows, chance does not introduce disorder into life; what we could call controlled probability enables new properties that would have been complicated to implement with nonprobabilistic systems.

Specialization and Modularity

Another simplex property of living organisms that we have identi-fied is specialization, or modularity. In building his phrenology of the cerebral cortex, German neuroanatomist Franz Gall ascribed a par-ticular function to each area. At the end of the twentieth century, this phrenology was criticized for segregating functions. In reality, func-tion is determined by networks of dynamically interconnected brain areas. However, each brain area performs very precise operations that can be of interest for many different functions. To take an ex-ample, language does not only involve Broca's area on the left side of the cerebral cortex, but we know that this part of the brain is essential because a lesion in it suppresses the capacity of speech. More gener-ally, a given area of the brain, or some of its parts, may participate in several networks because it is the locus of specific neural operations.

Faced with complexity, living organisms chose specialization, modularity, separation of functions, division of labor, categoriza-tion, and distinction. The brain is like a hive, an ant colony, a termite mound, an army, a factory, society itself. It is the seat of extreme specialization that appears very complex but that, in reality, simpli-fies processing of information about the world and, moreover, that facilities better control of action. Several examples will illustrate the extraordinary variety produced by the brain before it recreates the unity of perception and action.

To measure movements of the head, the vestibular sensors — the semicircular canals in the inner ear that measure head rotations—deconstruct movement of the head into rotations in three planes, forming a Euclidean frame of reference, corresponding to the planes of the canals. This geometric division in the same three planes also occurs at the level of the subcortical areas (the accessory optic system), which analyze the motion of the visual environment. A separate system (the otoliths) measures translations of the head and can also contribute to the three-dimensional encoding of head motion. Mathematicians have preserved the distinction between rotation and translation, which greatly simplifies the kinematic description of movements.

The visual system is another nice example of modularity. Ten areas of the visual cortex have been shown to encode different features of the visual world,[3] which testifies to the plural character of vision. But many of these areas turn out to be sites of multisensory convergence, all the while being organized in a manner that simplifies this convergence. I list below some of these visual pathways. Investigation into them confirms that choices have been made throughout evolution to modularize the analysis of the visual world in connection with action or function. Our aim here is not to give an extensive review of these multiple pathways but just to list them briefly (at the risk of sounding trivial for specialists) to point out the different types of analyses they perform.

A retino-collicular analyzer specialized for orienting gaze and body toward moving targets. Owing to a widespread research focus on the visual cortex, we tend to underestimate the role of the colliculus in perception. In fact, the colliculus plays a fundamental role in unconscious perception of movement and probably also of shapes. When playing a tennis tournament at Roland Garros (the French Open) and trying to return a service ball whose speed is 200 kilometers per hour, a player cannot rely on his cortical analyzers. Only the collicular route and anticipation will enable a successful return. In *The Brain's Sense of Movement*, I mentioned the case of blind children at a school in Montgeron who play Ping-Pong and basketball! They are classified as "blind," but their collicular vision of movement is probably intact, albeit not taken into account. The colliculus may not have the mechanisms available to the cortex for analyzing the details of the visual world, but it can very rapidly trigger adaptive behavior, in particular with regard to moving objects, as we have shown in work with neurophysiologist Alexej Grantyn.[4] In fact, even the activity of apparently "cortical" structures, such as the MT (the mediotemporal area, which encodes visual motion), can be activated by a rapid route going from the retina to the colliculus and to the MT through the pulvinar. Many recent experiments have confirmed this rapid route for processing motion. This, too, constitutes simplexity.

An analyzer that transmits images from the retina to the amygdala to quickly identify the positive or negative value of things seen (see chapter 3). We have already mentioned this pathway, which is part of the emotional (or limbic) brain. In less then 100 milliseconds, it can trigger an adaptive bodily reaction (fear, joy, flight, immobility). It can detect in this time window the presence of a snake and cause you to react. It is only later that the temporal lobe analyzes the shape of the snake and distinguishes a harmless garter snake from a viper. The amygdala also detects aggressive intention in the facial expression of others. When you meet someone for the first time, you immediately form an impression: "I like her," or "He frightens me." This first impression is the result of analysis carried out by the amygdala. This warning mechanism is clearly very ancient. It often must be inhibited by the orbitofrontal cortex. The advantage of this very fast analysis of the environment is that it alerts us to danger or, conversely, to the presence of a desirable agent. It is not simple; it is simplex.

An analyzer that corresponds to the dorsal pathway, specialized in processing the spatial localization of objects (the "where" pathway). This pathway begins in the parietal cortex and carries information toward and back from the frontal and prefrontal cortex. It also assembles information regarding the location of targets of interest. For example, it contains the "frontal eye field," which guides gaze movements, and the "supplementary eye field," which prepares gaze movements for directing attention to an object.[5] Other areas in the prefrontal cortex are involved in learning new sequences of gaze movements and in working memory.[6] This pathway is fundamental for attentional processes.

An analyzer that corresponds to a ventral pathway (the "what" pathway), specialized in identifying shapes of objects and living creatures. Along this pathway leading to the temporal lobe, groups of neurons are capable of identifying shapes; an area of the extrastriate visual cortex is even specialized in identifying the shape of the body (extrastriate body area).[7] Receptor fields — the windows through which neurons view the world — start small and increase in size through the visual pathways up to the stations of the temporal cortex, where faces, objects, and the environment are recognized. Groups of neurons in the temporal lobes have properties that enable them to respond in a selective way to innate and learned repertoires of elementary visual forms.[8] Sometimes, too, movement helps in identifying shapes, objects, and living creatures.

The late Marc Jeannerod, a neuropsychologist, proposed another pathway, *the "how" pathway,* which deals with the procedural aspects of grasping.

The brain even has specialized modules for identifying particular categories of shapes. A part of the fusiform gyrus is specialized in analyzing natural animal and human shapes, another in recognizing words, whereas the parahippocampus identifies the shapes that characterize the environment.[9] I think that this modularity or specialization facilitates analysis of the world and, more important, allows us to anticipate and to make predictions.

Most important and remarkable is that after all this modular decomposition, the brain—particularly the frontal and prefrontal cortices—recombines data from the dorsal and ventral pathways, adding appropriate information regarding context, memory, motivation, and emotion. Etienne Koechlin recently proposed an interesting model of these processes,[10] which he describes via an example. Three sections of the frontal and prefrontal cortices are involved in the decision to pick up an apple. One says, "I see an apple." The second says, "I like apples" (desire). The third says, "I will take the apple." Finally, a centrifugal (top-down) control mechanism for each of these components is exerted by the prefrontal cortex.

Speed

I stated earlier that in the primary visual areas, neurons can only see the world through very little windows called receptor fields, which measure about one degree in size. Contrary to what we used to think, these receptor fields are not fixed; they can change their properties remarkably fast: Evolution has endowed them with surprising properties that improve processing speed, thanks to the great flexibility of their properties. Visual cortex neurons in the primary visual cortex have been divided (by neuroscientists David Hubel and Torsten Wiesel, who shared in the Nobel Prize for Physiology or Medicine in 1981) in two categories (simple and complex) according to the type of stimuli to which they were sensitive. A "complex" neuron can be transformed into a "simple" neuron (again in the sense of Hubel and Wiesel) by modifying the amount of visual noise. Its response becomes more linearly related to visual stimulation. We speak of nonlinear systems when there is a complex relationship between a stimulus and a response. In technical parlance, simplifying things often means linearizing them.[11]

Another especially valuable mechanism is one that enables us to immediately detect a rapidly approaching object. The brain has found a very elegant solution to this problem: Instead of having to make complex distance calculations, all we have to do is to estimate the apparent surface of the object and its rate of dilation, that is, the relationship between its apparent surface on the retina and the speed of diminution of the image. The relationship between these two values, which are easily analyzable by the retina, directly gives the time to contact when the speed of the object is constant.[12] When the movement is variable, as I stated earlier, we use predictions based on internal Newtonian laws. Moreover, we can rely on the superior colliculus, whose role is to predict the trajectory of objects and to orient gaze toward things in the visual field. The colliculus identifies position, speed, and especially predicts the trajectory of an object. It triggers the movements required for us to orient ourselves in a predictive, anticipatory way. The colliculus allows a lizard to catch a fly, a tennis player to return a ball his adversary has served up at 200 kilometers per hour, and a goalie to block a ball during a penalty shot.[13]

Of course, signals relating to an obstacle, an enemy, or potential prey may not necessarily all reach the brain at the same time. This is the case of vision (the image is transmitted at the speed of light) and sound (which only travels at a rate of about 30 meters per second). Then what happens? To compensate for the temporal gap between the arrival of the two visual and acoustic signals, neural mechanisms (still mysterious) delay processing of one so it can be combined with the other, as if they were received simultaneously.[14] It is a splendid simplification of sensory cooperation.

Resolving Ambiguities

One of the greatest challenges for perception is how to resolve ambiguities resulting from errors caused by the limited properties of the senses relative to the complexity of the world. I believe that many illusions are actually solutions found by the brain to solve problems of perceptual ambiguity. They are decisions that simplify neuro-computation and enable action.

An example of a fundamental ambiguity is that which results from changes in the appearance of an object when we move. This altered appearance is evaluated by the brain in multiple ways. The most important, as French mathematician Henri Poincaré suggested,[15] appears to be information regarding the observer's own movement. When I perceive that an object has changed shape, if I am moving myself, it is likely that the change is due to my own displacement and not to a change in the shape of the object. This perception involves mechanisms that detect covariation between my movement and the deformation of objects.

A second example is the tendency of the brain to make objects or space symmetrical and its preference for rigid structures. These propensities enable the brain to resolve ambiguities contained in the optic flow by means of perceptual decisions — what I have elsewhere called the tyranny of perception. Arbitrating cases of conflict in favor of a solution that ensures the greatest stability allows the brain to complete the information provided by one of the senses with information furnished by the other senses. Since the otoliths in the inner ear cannot distinguish between acceleration in one direction and braking in another, vision determines whether the movement detected by these sensors is due to braking or acceleration.

A very special brain mechanism for simplifying perception permits the continuity of objects. Piaget devoted much effort to studying the development of this capacity of infants, once past a certain age, to continue to perceive a moving object even when it has disappeared behind a screen or been placed inside a box. Very young infants think that the object has ceased to exist. When they are older, they understand that the object that has reappeared on the other side of the screen or that is in the box is the one they saw earlier. This faculty is related to Albert Michotte's "phenomenal causality." This concept can be illustrated by a simple experiment. Consider two balls on a computer screen. Ball 1 is fixed and ball 2 moves toward ball 1. When ball 2 hits ball 1, we will give it a movement in the direction of the previous movement of ball 2. We then *perceive* that ball 2 has given ball 1 a kick and "caused" its movement. This phenomenon is based on an *inference*, which we can identify as a fundamental property of simplexity.

In humans, the capacity to perceive the continuity of the motion of an object in a cluttered environment is due to the ability to track an object in the absence of visual signals. When a moving object has disappeared behind a screen, the eyes continue to follow it even when it is gone. But how? The answer is an extraretinal signal (that is, information about the object's motion coming not from the retina but from the brain itself, which predicts the motion) sent to the visual areas, most probably at the medial superior temporal area in the temporal lobe, but also at the level of the visual cortex. Ventriloquism is another spectacular example of phenomenal causality. This elegant mechanism that governs complex processes is, in my opinion, typically simplex.

Separating Content from Context

An interesting way of simplifying the brain's analysis of the world for guiding action or decisions is distinguishing between content and context. Recall the two visual pathways mentioned earlier. One analyzes the location of an object or person — in other words, its context. This is the dorsal pathway, which goes from the visual cortex to the frontal cortex at the upper portion of the brain. The other pathway analyzes identity, or content (what? who?), along the ventral pathway that leads to the temporal cortex. An additional pathway that processes the emotional value of a situation via the amygdala is specialized for context. In other words, attributing qualities like "dangerous" or "beneficial" to an object or a person contextualizes the sensory perception. In the same way, we know that five circuit loops link the thalamus, the basal ganglia, and the cortex. These circuits select and control bodily movements, the eyes, and also memory and emotion. Each loop, which encodes sensorimotor content, is backed by another loop that encodes context.[16]

If, as neurophysiologist Rodolfo Llinás maintains in *I of the Vortex*, consciousness requires that content and context be adjusted by mechanisms that synchronize oscillations, how is it possible to simplify this compounded complexity? Let me suggest the following hypothesis: Separating context and content is important in extracting invariants in the world, independent of the particular circumstances of the perception or action under way. By the same token, perception of an object or an event that takes no account of context is useless. This essential separation may make it possible to dissociate learning about relationships between objects in the perceived world from learning about the context of action.

The Brain's Feeling for Shape

Having isolated objects either visually or by means of multisensory exploration, we can then turn to manipulating them. To do that, we must know something about their shape, solidity, and "compliance." The brain has available several simplex solutions to aid in these processes.

In feeling an object to test its shape or gauge its weight, or in caressing a person we desire, we not only call into play the tactile sense through the many receptors situated in and under the skin.[17] We also activate proprioceptive (sense of the body) sensors situated in the muscles and joints, and we associate that activity with a sense of muscular effort.[18] This complex ensemble, which requires elaborate multisensory integration, is what we call the *haptic sense*. This complex mixture of sense also exhibits simplex modes of processing.

In general, the brain exhibits a preference for gestalts,[19] shapes produced by a subject.[20] In other words, the brain gives physical forms in the world specific functions depending on their use. For instance, the stem of a flower is a ladder for an insect crawling up it, a handle for us in picking it up and offering it to our beloved, or a base for a spider to anchor its web to. The neural basis of this capacity to perceive shapes is not simple. It involves mechanisms that are very relevant to our purpose. For example, in contrast to what we once thought, encoding the shape of objects (or of a body) is not done uniquely through the firing frequency of the neurons that encode the objects' features (shape, color, luminescence, or sound). In fact, today it is believed that the temporal synchronization of the activity of different sensors produces the perception of unity of an object. The shaping of sensory signals in the thalamus by descending modulatory inputs is probably also related to this modification of perceived shapes according to our needs. Only a fraction of the information from the environment reaches the cerebral cortex. It has been proposed that only 5 percent of thalamic activity is produced by incoming sensory signals. The rest is from the brain itself!

The same is true of the simple *tactile perception* of the shape of an object using a finger. The simultaneity of firing of the excited receptors is what contributes primarily to the perception of the shape of the object. This temporal and spatial distribution of simultaneous activity at the precise moment of contact is apparently directly interpreted in the brain as a circle.

In addition, the brain also possesses anticipatory mechanisms. For example, the somatosensory cortex (S1 and S2), long considered to be an area of passive reception of information concerning a person's own body, contains predictive properties relating to tactile perception.[21] But this prediction is also facilitated by cooperation between vision and haptic perception. It has been shown recently that the areas activated by touch (S2) are partly the same as those activated by the sight of tactile contact.[22] When you watch a movie and you see one person touching another on the screen, you sometimes feel touched or kissed! An experiment was done in the brain scanner: Subjects whose head was in the scanner observed a subject's legs being caressed on a computer screen. Activity in the somatosensory area of the observer was recorded. Tactile and visual information interact at the level of the parietal cortex (the intraparietal sulcus),[23] and the haptic sense can improve the perception of rotation caused by optic flow.[24] At the level of the thalamus, too, certain signals carry out predictions. We know that the traditional role attributed to the sensory thalamus is to passively transmit information flowing to the cortex. However, if sensory stimulation is associated with a reward, the thalamic neurons link the sensation to the reward. A phasic response has been observed to appear very early after the stimulus, depending on the sensory modality. We call this retrospective encoding. A delayed, particularly interesting response then occurs and is greatest just before the reward. This second response, independent of the sensory modality and modulated by the value of the reward, predicts the reward to come. We call it prospective encoding.[25]

The Simplexity of the Haptic Sense: How Not to Crush a Raspberry

Suppose you wish to grasp a raspberry between two fingers without crushing it. The force you need to exert to not crush the fruit must be very finely regulated. It must be adapted to the compliance of the raspberry, that is, to its elasticity (compliance is the opposite of stiffness). Because this cannot be known a priori, before contact, the risk of crushing is great unless the brain can predict the parameter.[26] It so happens that the brain possesses an internal model of how the fingers work and develops an internal model of the compliance of the berry. Regulation of the force thus becomes a simple comparison between the estimated value (the compliance of the raspberry alone) and a value measured at the moment you grasp the fruit, thanks to proprioception, that is, the properties of the sensors that detect displacement, pressure, and force in your hand. This ability to anticipate compliance saves you from crushing the fruit. This anticipation is not simple, but it clearly enables simplexity.

The brain can inform us about the shape of an object based on motor activity alone, or it can extract kinesthetic information based solely on information from the skin.[27] It can also perceive the form of an object without any proprioceptive, kinesthetic, or motor information. In fact, human vision is capable of inducing a perception of haptic force all by itself. Consider the following example. The image of a virtual piston acting on a spring is shown on a computer screen, and the viewer has the possibility of pushing the piston down by moving the computer mouse. If the piston moves on the screen as one would expect with a spring that is slowing down, the person holding the mouse will feel a force being exerted on the mouse. This illusion, called "pseudo-haptic illusion," was discovered by computer scientist Anatole Lécuyer. A single piece of visual information induces the illusion of a force applied to the hand. Philosopher Maurice Merleau-Ponty, who famously insisted on the role of perception in understanding and engaging with the world, was right when he said that vision is the brain's way of touching.

Why Is the Tip of a Finger Round?

Every simplifying principle in biology, no matter how basic, represents compelling ideas and is potentially applicable to other areas of thought. Vincent Hayward,[28] a colleague who specializes in haptic perception, asked me one day why the tips of our fingers are round. He concluded that whenever we press a finger on an object, the area bounded by the contact forms a circle. The tip's special geometry facilitates extraction of variations in pressure on the finger, which provides us with information. It's nonobvious, but simplex!

Here is another example: One of the mysteries of the hand's anatomy is the delicate ridges of the skin of the fingers, which serve to identify each of us uniquely. Recently, the existence of these patterns was explained as enabling the brain to carry out a "wavelet" analysis, a type of mathematical analysis that extracts the dynamic characteristics of very diverse processes. Generally speaking, the perception of movement is a powerful means of identifying objects and resolving ambiguities. For example, the movement of an object traveling alone on the skin is perceived as an insect walking on the skin, or in any event, an animal capable of autonomous locomotion. Psychologist David Premack has proposed that autonomy of movement be considered a criterion for life.

Detecting Smells: Dimensionality

For many animals, smell is the most important sense because it enables recognition of prey and predators, aids in route planning, and guides navigation. In vertebrates, smells—which are generally subtle combinations of primitive aromas—are processed by a group of neurons concentrated in the olfactory bulb. This neuronal architecture ensures identification, categorization, and memorization. The neurons of the olfactory bulb, which are organized in an extraordinarily complex network, differ by their excitatory and inhibitory properties, their shapes, and the branching of their dendritic tree. The ensembles tend to oscillate in time, between 20 and 30 Hz (cycles) per second.[29] When they are recorded, the oscillations reveal very elegant solutions to this complex problem. When presented with an odor, every one of the neurons of the olfactory bulb reacts differently, and fluctuations in oscillatory activity appear in the network. When the network oscillates at a given frequency, the temporal signal is strongly sinusoidal, consisting of troughs (negative) and peaks (positive). For a specific odor, certain neurons are active and produce an action potential at precise moments of network oscillation.[30] In-depth analysis of the firing of these neurons shows that the olfactory bulb carries out some sort of classifying process according to the category of odor.[31] In this way, the network is multiplexing: It simultaneously transmits different types of information in the same flow of data.

In the language of mathematics, these types of networks *reduce dimensionality;* that is, they simplify the space in which the activity of the complex ensembles is represented by reducing the number of dimensions. Consequently, in this three-dimensional space, processing can be deconstructed into a trajectory comprising three steps: Initial zero activity (before the odor) serves as the starting point, that is, the coordinates (o, o, o); in the second following presentation of the odor, the trajectory reaches the maximum distance from the origin; with the odor still present, the trajectory is stabilized at a certain point, different from the origin, called the fixed point. Studies show that, although trajectories differ depending on the odor, they are identical for any given odor. This type of encoding de-multiplexes the quantity of potentially memorizable information.[32] The maximum discrimination between odors is achieved at the maximum of the trajectory. This system evolved to cope with ecological conditions in which odors are often only very briefly present. It is a fine example of simplexity because it possesses complex properties that simplify neurocomputation.

Making a Simplex Decision

Another example of reduced dimensionality is that of the brain mechanisms involved in decision making. Suppose that you are going shopping and you arrive at an intersection. To the right is the road to the post office, where you need to mail an urgent letter. To the left is the bakery. If you go left first, you could treat yourself to a nice piece of pastry, but it's late in the afternoon and stopping there could make you too late for the post office. This type of decision making was reproduced with a rat in a T-shaped maze. According to the position of an illuminating light, the animal had to choose between the two arms of the T at the ends of which it might receive a reward. Recordings of the activity of the neurons of the rat's prefrontal cortex, where such decisions are made, showed that populations of neurons were activated at the same time; in other words, they were synchronized. These assemblies were formed at the moment where the rat became convinced it would receive a reward and decided to go left or right. Moreover, these same assemblies re-form during sleep, when the patterns of neuronal activity that resulted in successful behavior are consolidated.[33] Here, simplexity has to do with the mode of encoding, which reduces the dimension of the problem through temporally synchronizing activity. Time is a friend of simplexity, as we will see in greater detail in chapter 10.

Reduction of dimensionality and temporal synchronization are some of the many mechanisms that simplify perception by borrowing pathways that correspond to the criteria of simplexity defined here. Faced with the complexity of the world, life has found solutions of exceptional elegance, at the cost of apparent detours that themselves contain a certain dose of complexity. Despite these detours, the solutions facilitate elaboration of perceptions that are guided by the intention of the subject. They allow anticipation, flexibility, and speed to react adequately to sudden events on which, for example, survival depends. Paradoxically, this multiplication of modes of encoding actually enhances efficiency by simplifying the process of neural calculation. Of course, all of these qualities have a price. Sometimes, as with critical periods, it is the existence of a very precise calendar, where the individual has only one chance; at other times, the price is limited perception. But, as we like to say, "There is no free lunch." Simplexity always has a price, although the return on investment is great.

Part II. WALKING ON THE MOON

7. The Laws of Natural Movement

Have you ever seen an anatomical cutaway—one of those images that reveal the extraordinary complexity of the network of subcutaneous nerves that link the muscles to the spinal cord and ultimately to the brain? One of the problems posed by this complex network is that of transmission delays. In the human male, the distance from the foot to the cerebellum is around 1.80 meters, whereas a distance of barely 15 centimeters separates the neck from the cerebellum. As a result, if the speed of transmission along the nerves was the same for the feet and the neck, it would be impossible for information regarding events occurring in these two regions to reach the brain simultaneously. Furthermore, synchronization of movements of the two portions of the body would be very difficult. Here, too, a solution exists: The speed of transmission is not the same. It is faster for the feet. This solution is simplex because it involves a detour through an accrued complexity: regulation of various transmission speeds adjusted for distance. As I will show in the following pages, the organization of motricity includes a variety of other simplifying principles. Some of these principles can be called "laws of natural movement." Not only do they take into account kinematic and dynamic solutions that optimize the performance of movement; they are also embedded in the physics of our planetary environment. For example, they take into account, and even use, gravity. If we really mean one day to walk and to live on the moon and on Mars, we will have to adapt these "laws of natural movement" to the particular values of gravity on other planets!

An Essential Simplification

In *The Brain's Sense of Movement*, I touched upon the problem of reducing the number of degrees of freedom and simplifying laws of movement. Let us revisit one of the examples from that book. Suppose that you extend your arm and use your finger to trace an elliptical shape. You will note that the velocity of the movement varies along the trajectory—we call that "tangential velocity." Your arm moves faster through the areas of lesser curvature. A very strict linear law, known as the *one-third power law,* links tangential velocity and curvature. The origin of this law is still a matter of debate. It surely is related to principles of energy minimization, jerk, and optimizing control. It describes simplex processes. But the remarkable thing is that psychobiologists Paolo Viviani and Natale Stucchi together made the important discovery that this law also governs the perception of natural movement.[1] For example, if you trace with your finger an ellipse in space and ask somebody watching you if your movement is at constant velocity, the other person will only perceive your finger movement as being at constant velocity if, in fact, you vary the velocity according to the one-third power law! These findings have an important field of application in the new world of animation with computer graphics used in digital movies and in the educational game industry: If the experts in these areas wish to create artificial personalities (avatars, say, or humanoids) whose movements should appear real to viewers, or at least as close to living creatures as possible, the movements of their creatures must obey this law. The same will be true for humanoid robots used for companionship; the movements of their limbs or legs will have to obey the laws if they are to appear "natural."

Now, instead of tracing an ellipse, make more complicated movements, like those illustrated in figure 7. You will note that your hand and arm are twisting in ways that make the movement complex. A very close link exists between the velocity along a trajectory, curvature, and torsion. Velocity depends on the torsion in a one-sixth power relationship.

We recently showed that these laws of natural motion probably derive from the fact that the brain works in accordance with non-Euclidean, affine geometry, as suggested by Jan Koenderink, for visual perception. Here, simple laws reveal that the brain uses multiple processes and several geometries (see chapter 10).

The Secret of the Nevers Attack

Locomotion, postural control, grasping and capture, avoidance, flight, and many other behaviors are performed by means of innate motor synergies that are both specific to each species yet also universal. Ever since Nikolai Bernstein began studying the movements of humans at work in the 1920s, identifying these synergies has been the focus of intense collaborative efforts by physiologists, engineers, and mathematicians. Today the search is on particularly for what are called "primitive motor action complexes,"[2] that is, synergies whose combination induces a complex movement. These combinations of primitives belong to a universal repertoire of movements that are partially innate and partially acquired during infancy.

You need only watch a dancer to realize that the repertoire of these synergies is limited but that they can also be learned by training. The stereotypical character of motor synergies and the fact that our system of mirror neurons enables us to more easily simulate familiar

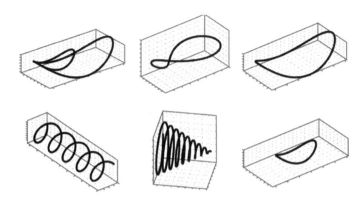

Figure 7. Examples of hand movements in three dimensions. It is known that for planar movements of a hand in space during drawing, for instance, there is a tight relation between tangential velocity and curvature (see text regarding the one-third power law). Here we show examples of tracings of three-dimensional hand movements in space, which add a component of torsion. It has been shown that these movements follow a simplifying law that links tangential velocity along the trajectory, curvature, and torsion: $v = \alpha \kappa 1/3 \cdot |\tau| 1/6 \cdot v$, where v is the tangential velocity, α is a coefficient, κ is curve, and τ torsion. In order to "look" natural and be easily decoded by the brain of an observer, the movement of an artificial artifact like a robot will have to obey this law. [Adapted from U. Maoz, A. Berthoz, and T. Flash, "Complex Unconstrained Three-Dimensional Hand Movement and Constant Equi-Affine Speed," *Journal of Neurophysiology*, 101 (2009): 1002–1015. See also D. Bennequin, D. Fuchs, A. Berthoz, and T. Flash, "Movement Timing and Invariance Arise from Several Geometries," *PloS Computational Biology*, 5 (2009): e1000426.]

movements also allows us to surprise others by breaking these rules and laws.[3] For example, the famous "Nevers attack," a sure way of knocking off a fencing opponent, must have been a new, nonpredictable move at the time — at least for the unalerted. I can also imagine that, in the martial arts, some gestures belong to a nonnatural repertoire derived from the novel combination of synergies or motor primitives, atypical movements acquired by virtue of arduous training, which can create new synergies. Some of the Eastern martial arts use clever combinations of movements from a common repertoire of "counterintuitive" gestures and movements. Similarly, comedians like Jacques Tati, Buster Keaton, and Charlie Chaplin made people laugh by surprising them with impossible or counterintuitive movements.

Great effort has been expended by physiologists to try to identify these basic synergies, which are often called "primitives." The methods used involve statistical calculations such as "principal components analysis," independent components analysis, and so forth. The aim of these statistical analyses is to reduce the kinematic variables (such as position and velocity) of the limb motions or patterns of muscle activity to a few basic variations. Results show that the apparently complex limb movements or muscular activities of the hundreds of muscles can, in fact, be reduced to a few simpler "components" called "principal components." These components are thought to reflect basic muscular "programs," synergies, or primitives that, in combination, construct movement. Thus, four or five principal components have been identified for locomotion and manipulation of objects, and three components for the bodily expression of emotion.

However, this approach does not really shed much light on the fundamental elements of the physiological repertoire. Cognitive scientist Emanuel Todorov, who studies movement control, writes:

"Little independent evidence exists that the principal components correspond to anything real. The presumed function of motor primitives is to simplify a complex control problem by reducing the dimensionality of the space where control solutions are sought. What simplifying assumptions are safe to make for all tasks that need to be performed in the future?"[4]

When Anatomy Dictates Geometry

Say you want to build a humanoid robot and want to be able to move its eyes or limbs, like a marionette. For example, you decide to have it make an oblique movement of its eye. But your robot only has two motors, one for moving the eye horizontally and the other, vertically. To make an oblique eye movement, you will have to coordinate the two motors. It is not easy! In chapter 3, on gaze, we saw that the problem of noncommutativity of rotations is solved anatomically by having the muscles of the eyes use pulleys. But other simplex properties contribute to this process. For instance, the brain has three "motor networks" for making visual saccades. The motor networks consist of three neuronal networks called "saccade generators," which are situated in the pons and mesencephalic reticular formation. They enable the eye to turn in three planes (horizontal, vertical, and torsional). The massive simplification devised by evolution consists in ensuring coordination by certain neurons in a subcortical structure called the superior colliculus, which projects to the three generators. The ingenuity here is the "branching" of the neurons, that is, the projection of their axons, which controls the direction of saccades. Thus, a neuron projecting to the horizontal generator results in a horizontal saccade, whereas one projecting simultaneously to the horizontal and vertical generators produces an oblique saccade. The "synergy" is determined by the anatomy, since the geometry of the movement is dictated by the axonal branching.[5]

In the nervous system, control of the geometry of movement by axonal branching is also exerted on actions of the limbs. In tracing an ellipse, we call into play a dozen arm muscles that determine the trajectory of the finger. The brain does not control each muscle separately. As mentioned above, it controls groups of muscles by virtue of coactivations—synergies. These synergies are controlled by the neurons of the pyramidal cortex (the motor cortex), whose axons project to very special groups of neurons, each of which produces a specific synergy (rotation of the wrist, extension of the arm, and so forth). The brain thus is able to simplify commands because all that is required to move the wrist is to activate a pyramidal neuron. The anatomy of command neurons ensures coordination and enables combination and orchestration of synergies into more complex movements by reducing the dimension of command space. This is an elegant form of simplexity.

Internal Models

I have said elsewhere that our thoughts, that is, our highest and even most abstract cognitive functions, are rooted in bodily action, and that the brain has developed the ability to anticipate the consequences of action, projecting onto the world its preperceptions, hypotheses, and interpretive schemas. For example, if you are grabbing a glass of wine in a hurry, your brain must calculate the position of the glass with respect to the table supporting the glass. It must calculate the position of your hand relative to your shoulder, your shoulder relative to the chair in which you may be sitting, and the chair vis-à-vis the table. Only then can the brain program movement by reckoning the force it will have to transmit to each of the 20 or 30 muscles involved. Once you have grasped the glass, it will have to take into account its weight, and so forth.

This process is immensely complex, and roboticists have come up with many solutions for simplifying the task, such as encoding movement by including only the relative distance between the object and hand. This kind of encoding avoids having to calculate changes in coordinates. But as a solution it is insufficient, because as the hand approaches the glass, it must be prepared to grasp the shape of the object. There is a simplex solution for doing that: The hand anticipates holding the glass to adapt to its shape. But this method is not efficient, because it gets in the way of speed. To be able to grasp rapidly, the hand must take the form of the glass before contact. To do that, evolution has invented a solution: The sight of the object is processed in the brain by two separate pathways — the dorsal pathway, which encodes the position of the glass in space, and the temporal pathway, which encodes the shape of the glass, identifies it, and combines this information with the intended purpose in the frontal cortex (see chapter 6). This functional segregation appears to complicate processing, but it has advantages. As soon as the hand moves, vision — which has identified the shape of the object and its use (the intended purpose) in the temporal lobe circuit[6] — preforms the hand in advance to give it the shape of the object, and this anticipation prepares the act of grasping. Thus, at the cost of a little "division of labor" and a second segregation — this time between the control pathways of the arm and those of the hand — the brain manages to achieve this simplex operation.

But the task of grasping is not always this "simple." If, instead of picking up a glass, you wish to catch a falling bottle of water or take hold of a heavy object, you must also anticipate the weight of the object or the force with which it will strike. Shepard was the first to hypothesize that the nervous system internalizes the physical properties of the world and that this internalization delimits our perception. Today, we have the proof of this internalization of the physical world, in particular Newtonian laws on acceleration of the body under the influence of gravity.[7] When you catch a falling ball, the brain anticipates the force of impact of the ball on the hand, which indicates an internalization of the effect of acceleration on the ball. One can generalize this finding to many other behaviors.

Generally speaking, a simplex solution to the problem of control of movement consists in constructing in the brain *internal models* of the body and the laws of the physical environment.[8] These internal models are neural networks. Very briefly, sending a signal to one of these networks to move the arm 30 degrees, for example, generates an output signal that takes into account the biomechanical properties of the arm as if the command had been sent to the actual arm. We conclude that the cerebellum possesses "internal models" of the dynamics of the limbs, models that enable the brain to simulate the trajectory of a movement before executing it.[9] The cerebellum, together with some areas of the cerebral cortex and possibly basal ganglia, also participates in creating internal models of the laws of the physical world.

These internal models can be modified by learning. In humans, the neocerebellum, which appeared very late in the course of evolution, may also contain internal models of objects in the world. An experiment was conducted to test this hypothesis.[10] When subjects placed in a functional magnetic resonance imager are taught to use a new tool, such as a computer mouse, through modified conditioning — when the mouse is moved from left to right, a target on a screen moves from bottom to top — two types of activities are recorded in the brain. One activates the entire cerebellum and depends on the amount of motor error during learning. The other is limited to a posterior part of the cerebellum and persists after learning. Most probably, this latter activity corresponds to the construction of an internal model of the tool with its new laws of functioning. Note, however, that this concept of an internal model was recently refuted by recordings of Purkinje neurons in the cerebellum. In fact, the results indicated that during movement, these neurons encode the kinematics of the movement and not, as roboticists maintain, an inverse model of the movement.[11] The debate is still open.

Changing Variables to Reduce Complexity

An important source of complexity in nature and for living organisms is *nonlinearity*. In a nonlinear system, behavior is not related in a simple way to a perturbation or stimulation. For movement, nonlinearities result in complex properties of the self-organizing character of the brain, that is, it does not depend uniquely on the effects of the environment, but on force, energy, and internal modifications.

In the face of nonlinearity, what can we do? One possibility is to process data by changing variables. Say you would like to guide a movement of the hand to track a target; in this case, position would be the variable to control. Solving this problem requires nonlinear equations so complicated that roboticists have proposed ways of getting around them.[12] For instance, the brain may use "composite" variables, that is, combinations — of position, speed, and acceleration. Proceeding in this way, instead of becoming more complicated, the problem is rendered "linear." Composite neural activity has been found in several structures that control the movements of the eyes and the hands.[13]

The nervous system may also simplify calculations by transforming highly nonlinear operations into a series of "generic" nonlinearities followed by a linear operation. The formal networks that function in this way include the most basic ones.[14] Here, nonlinear functions are deconstructed into weighted sums of nonlinear functions (sinusoidal for Fourier transforms, Gaussian for radial basis function networks). Such an approach reduces learning to a linear problem.

Consequently, it suffices to compute the weighted sum. To do that, we have very simple rules, such as the "delta rule," [15] which adjusts weight in a perceptron-type artificial neural network.

Uncertainty and Probability: Bet Rather Than Calculate

One of the yet-unsolved questions of perception is What are the mechanisms by which the brain creates a unified perception from the great variety of sensory inputs and the remarkably different reference frames in which it senses the world or the body in action? It is obvious that, in fact, all sensors operate with a great degree of uncertainty. Integration of sensory information — multisensory integration — is therefore an example of complex processes that can be approached through probabilistic quantitative methods based on Bayesian inference,[16] as we saw earlier. The idea of probabilistic Bayesian reasoning is to keep to the essentials, to what is "most often true." For example, a small, elongated object that is moving slowly is probably an earthworm, and a toad will snap to it even if it is a piece of wood! A Bayesian observer who is interested in earthworms will attribute a fairly high probability to the proposition "This is a worm," associated with the arrangement of clues — "small, elongated object that is moving slowly." It is simpler than constructing an exhaustive list of everything that might resemble a worm, associated with the sets of clues that are indispensable for determining, without a doubt, what the object is. Bayesian models (see chapter 2) simplify representational systems and research into sets of clues that are relevant for describing a complex multisensory reality.[17] In addition to the many behavioral studies that justify interest in the Bayesian approach, several efforts have reinforced the idea that the brain estimates probabilities.[18]

In general, the calculations required to integrate sensory data are nonlinear. But these calculations may be reduced to linear operations inasmuch as the probability distributions of the variables for sensory data are themselves essentially linear. This also makes it possible to estimate "maximum likelihood" — a term that captures how faithfully a value represents what one wants to measure — using locally linear attractor networks.[19] This approach reduces perceptual decision making to simple linear operations. Unfortunately, not all Bayesian inferences can be handled in this way. Such is the case in implementing Kalman filters (an effective method of dealing with "noisy" measurements, named for the engineer who invented it)[20] or changing coordinates. It is never easy to measure the simplicity of a calculation, even if the brain is able to "calculate."[21]

Contraction

One of the greatest challenges of neuroscience today has to do with the brain being composed of thousands of circuits, each with feedback (the information is fed back to where it comes from) forming "loops" that make up "functional networks."[22] This organization results in immense complexity, but my hypothesis remains the same: A few simplex principles enable cooperation of these networks. Jean-Jacques Slotine has shown that "contraction" is one of the properties that can aid coordination.[23]

In brief, a system is said to be contracting if it has the following characteristics. When it is shifted from its trajectory (a kinematic trajectory, such as a gesture or a change of state), it returns exponentially fast, independent of the initial conditions. Moreover, a contracting system can be included in other systems, themselves contracting; the resulting group is also contracting, which gives it several interesting properties. Slotine showed that all it takes to stop oscillations in a contracting system that contains many networks is a single inhibitory neuron.[24] Now, the brain is meticulously made of numerous systems that must be rapidly activated and deactivated. We have applied this theory to controlled systems in the basal ganglia. Recall that five systems linking the thalamus, the basal ganglia, and the cerebral cortex have important roles, especially in choosing action. They have been modeled many times over. However, we have been able to show that contraction contributes interesting combinatory properties.[25]

8. The Simplex Gesture

Animal and human gestures are used for action but also as signposts, symbols, and to indicate intention. They are neither simple nor complex. They are simplex because, in a very global and immediate way, they enable the brain to grasp a reality, an emotion, a thought, or a complex social relationship. Gesture is a fundamental mark of culture and art, which are always simplex expressions. Gesture is essential to them. Drawing, painting, music, mime, acting, sculpture, and dance are always expressed in gestures. For this reason, we must not be content with a physiology (or a philosophy) of action. We need a physiology of gesture, of bodily expression, and of intersubjectivity.[1]

The Concept of Gesture

Let us begin by clarifying the concept of gesture. For the Greeks, gesture (action) is one of the basic components of eloquence. Gesture crops up repeatedly in the writings of Aristotle, of Cicero, and of their successors in the Middle Ages. *Actio* was intended to express the movements of the soul in three domains: physiognomy (*vultus*), voice (*sonus*), and gesture or movement (*gestus*), to which I would add bearing (*incessus*). I think the same is true of discourse. The lectures I give at the Collège de France, or in front of audiences all over the world, have led me to work on eloquence. As a result, I am convinced that expository speaking naturally involves gesture: A lecture is a choreographic performance.

The Middle Ages abound in expressions that are not limited to language. The Lithuanian symbolist poet and translator Jurgis Baltrušaitis gave us a medieval iconography, but he downplayed a fundamental aspect: the iconography of gesture.[2] The study of gesture in the Middle Ages is edifying. It derives from ancient norms. Gesture is more than movement (*motus*).[3] It is basic to our relationships with others. It simulates actions—acts—better than words ever could. It can be incomplete; it may be a sign, or a symbol (*signus*); it is immediately comprehensible; it triggers a "mirror" effect; it is also one of the components of *habitus*.[4]

Gestures range from useless—*gesticulatio*—to subtle, and to suspended forms: For example, the pause before an image, often accompanied by a stereotypical posture. A well-known example is that of the "biomechanics" of the great Russian director Vsevolod Meyerhold.[5] The exercises that Meyerhold demanded of his actors deconstructed actions, such as aiming a blow (see fig. 8) or walking, according to very precise rules intended to give them total control of their bodies. Thanks to Béatrice Picon-Vallin, a specialist in Russian theater, we had the opportunity to record some of these exercises with one of Meyerhold's last living students, the director Alexei Levinski, of Moscow. Figure 9 shows the recordings of a stone thrower broken down into three stages: movement, known as *stoika;* and two phases, called *otkaz* and *posyl*. More precisely, *stoika* represents a "suspended movement" or a "movement within immobility." (Immobility is important: We find it in the instant where everything stops and everything is said in Japanese *ma* or in the musical sigh, of which pianist Glenn Gould was a great specialist.) *Otkaz* and *posyl*, however, are associated with active movements. Furthermore, *otkaz*

Figure 8. Exercise in deconstructing movement. Russian director Vsevolod Meyerhold's enactment of a blow to the face with the hand. The photos were taken on a roof in Moscow. In these types of images, Meyerhold was looking for the successive phases of the movement (see text). [Courtesy of Archives of Béatrice Picon-Vallin; see also V. Meyerhold, *Écrits sur le théâtre*, vol. 2 (Lausanne, Switzerland: Éditions L'Âge d'Homme, 2009).]

represents a movement opposite to the following *posyl*, and probably is intended to emphasize what is to come.

I see gestures as falling into three major categories. Gesture can be a simple movement: Reaching for a cup of tea is a simple gesture. Gesture can also be a mode of simplified encoding: Think of the gestures of sailors who, prior to Morse code and long-distance wireless, communicated with flags following a very rudimentary code. The military salute, army protocol, and the antics of a policeman at an intersection—these are all conventional expressions of hierarchical rules. Finally, gesture can be a sign of emotion, of intention, of re-

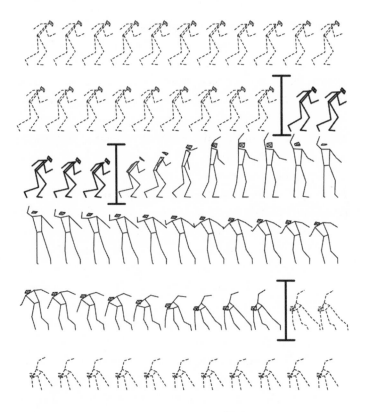

Figure 9. Deconstruction of different phases of movement of an actor pretending to throw a stone, according to Meyerhold's notions of biomechanics. The gesture is broken down into three types of movements: *stoika* movement (shown at top and bottom), the *otkas* phase (shown in the second section), and a phase called *posyl* (shown in the third section). ["Vicon" Motion System recording courtesy of Moscow actor and director Alexei Levinski. Adapted from Y. Carvalho, B. Picon-Vallin, A. Terekhov, and A. Berthoz, unpublished paper; see also the thesis of Yedda Carvalho, "The Biomechanics of V. Meyerhold," University of Paris, 2008. Courtesy of A. Levinski.]

gret; it may also have an abstract meaning. A mathematician friend of mine who was participating in one of my courses at the Collège de France suggested that a simple gesture sometimes explains a geometrical idea better than an equation. In fact, this last category of gestures is not simple but simplex: They make it possible to communicate feelings or complex ideas through a remarkable shortcut—a movement of the hand, of the fingers, and the body. This is what gives sign language its power.

The fact of communicating by gesture implies the other in one's own body. The auditor becomes agent; the spectator becomes actor. Gesture may be immobile. A magnificent example of gesture that is both rigidly encoded and also a source of infinite variation is the Renaissance paintings of the Annunciation.[6] In these images, Mary receives the message from an angel, who is generally positioned to the right of the image. The Holy Spirit, represented by a dove, dominates the scene, and light sent by God above provides illumination. Very often, a column separates the angel from the Virgin, pictured in her bedroom. The scenes follow a strict template, yet the painters were free to give the Virgin an attitude of their choosing, and especially to *choose a gesture to express varied sentiments—surprise, acceptance, submission, hesitation.* This gesture is what gives the painting its meaning. Here, simplexity derives from a combination of very strict rules of composition and infinite variations that individualize the painter's message and the observer's interpretation. I have always been struck by the total absence of emotions in Italian Renaissance painting. The faces and bodies radiate a serenity that I used to attribute to a desire to make the faithful feel at peace with their souls. Yet when one looks more closely at the gestures of the hands and the postures, one sees the categories described by Barbara Pasquinelli, a specialist in gesture: expressive gestures, gestures of despair, obscene gestures, and rituals.[7] For a long time, I could not figure out the code.

Sometimes simplexity implies that the brain possesses information—for want of a better word—for giving meaning to gestures. This information does not come only from the external world; it demands a certain resonance between the type of information and the laws governing the brain's functioning and interpretation. Just as I cannot understand a language that I do not speak, I cannot understand the *Noh* of Kabuki if I do not possess the necessary mental models. Simplexity frequently requires learning. Because we have a repertoire of genetically encoded shapes in the amygdala for identifying menacing or aggressive gestures, we do not have to learn them; on the other hand, we do need to learn the meaning of gestures connected with human relations, with social norms, and with religious symbols.

Matching Gestures and Words

Much has been made of the role of language in communication. There is no doubt that access to language enables the rich exchanges that make humans different from other animals. Moreover, language is a mechanism of complexity because it allows simulation of reality by expressing it in signs and symbols. The universal character of the laws that underpin language is well established, and these rules make it easier for us to understand others. At the same time, the diversity of languages also allows each population and each culture to express what is unique about them. For all that, other modes of communication with extraordinary potential are often underestimated. Humans, like animals, possess a nonverbal mode of communication that is based on gesture and bodily expression. Here, too, there is a duality between the diversity of gestural language and its cross-cultural invariants. The repertoire of gestures is vast,[8] and is always accompanied by postures and attitudes.[9] They are not only the expression of a body, but of a person:[10] They have a context.

Today, you have only to stop outside a school filled with students, or to watch a politician on television, or a singer, or a doctor to realize that the repertoire of human gestures is constantly being adapted to skills, professions, and so forth and that it is adapted to the vicissitudes of human relations. Just observe people using their cellular telephones on the street. I did this experiment in Italy, where gestures are especially expressive. In Pisa, a young woman was pacing to and fro in the market square, arguing energetically. I had no idea what her conversation was about, because she was too far away. Yet I was profoundly affected by the message of her gestures, these attitudes that accompany words. I still feel emotionally overwhelmed just thinking about it. This is a perfect example of the facts that gesture accompanies thought and shapes it and that it can capture the complexity of a situation. Gesture is a manifestation of simplexity because it is an immediately comprehensible statement of a complex reality. *It contains the essence of an act, not only an action.* It reflects purpose and content. It takes into account the state and identity of the person who is gesturing. It also anticipates future action. We perceive gestures "immediately." A sequence of gestures bears us along like a fragile canoe on rushing waters. Gifted orators know full well the impact of using and abusing gesture. As the great roman statesman Cicero described presciently in his treatise on eloquence, gesture is evidence.[11] Simplexity is evidence, too.

Work Gestures

One of my drawing teachers at the Lycée Saint-Louis used to say, "Make it look effortless." If we were drawing a face, for example, he would make us redo the suggestion of a mouth a hundred times. In every profession where gesture is essential—surgery, construction, lute making, cooking, design, hairdressing, car repair, piloting, music — workers know the importance of apprenticing. Anyone who has practiced a manual skill for a long time knows that the gesture becomes more expert, rapid, and precise and that it develops a personal character. The initial hesitations, futile detours, and false starts give way to serene assurance and adaptability. Halting efforts are replaced by an often unique gesture characterized by few false starts and a fluid chaining of movements, as in a melody. At this stage, the gesture moves toward its end instead of being conceived as a number of steps, a "voyage" rather than a series of paths and actions. I myself had such an apprenticeship during a period of several years when I was conducting surgical experiments to elucidate the neural connectivity of all the muscles of the eyes. The operations to isolate the oculomotor nerves lasted between six and eight hours. Eventually, over time, I was able to connect the fine, precise, and rapid gestures — avoiding any unnecessary motions — that revealed the astonishing anatomy of the oculomotor apparatus. Gesture joined seamlessly with the beauty of the living organism by following its forms and respecting its vessels. In some way, through the movements of my fingers and hands, but also of my posture and gaze, I became a part of the organ that I was studying. Suddenly, surgery seemed simple to me; in actuality, it had become simplex. "Effortless action" (*wu-wei*) is a concept that one finds in Chinese philosophy. It has been analyzed by Asia scholar Edward Slingerland.[12]

This is the origin of the gesture that makes dance out of movement, and of the joy that produces the characteristic features of a drawing by Leonardo da Vinci or Rembrandt. Each curve is a gesture; every one of the strokes that shadows the shapes of the bodies and faces evokes the curve that a finger makes in tracing the skin of an object of desire in a soft caress. The philosopher and psychologist Théodule Ribot emphasized that emotion is first "e-motion," in other words, movement.[13] At core, gesture represents the very essence of the simplexity of life.

The Sacred Gesture: Jesus and Buddha

Every human manifestation of the sacred has used gesture to establish the sharing of emotions, ideas, and moral codes.[14] Sacred gestures—as they are practiced, say, during the Catholic mass—make this very evident.[15] The institution of the mass is traditionally organized into four parts: people, things, words, and actions. The latter consists of acts, movements, and gestures. The priest accomplishes three main categories of gestures during the mass. There are four "movements" (*motus*)—"left to right," "right to left," "altar to chair," "chair to altar"—and four "gestures" (*gestus*)—"extending the hands" (*extendo manus*), "raising the eyes" (*elevando oculos*), "bending the head" (*humiliando caput*), and "bending the body" (*inclinando corpus*). This repertoire also includes eight "action" gestures (*actus*)—"washing," "receiving," "placing," "swinging the thurible" (which contains incense), "making the sign of the cross," "lifting," "taking," and "breaking." *Each gesture is more than a sign. It is an act that is perceived, imagined, and lived.* Because every gesture seen by the faithful is also simulated in the brain, thanks to the system of mirror neurons, gesture is fundamentally much more efficient than words, and religious have learned to marry gestures and words to increase their evocative power.

The posture and gesture of Jesus on the cross constitute another good example of this evocative power of gesture. The open arms express both the weight of the martyred, suffering body, the gift of self, the call to others, and tenderness, like the combined gesture of a child seeking the extended arms of its mother. This attitude of tenderness and welcome has been studied by the child psychiatrist Julian de Ajuriaguerra. In Jesus, the posture expresses both despair and hope.

In addition to the arms, the bent head of Jesus shows a mixture of submission, resignation, and sadness; when lifted, it can also express the desire to resist and to fight. Sculpture allows a considerable repertoire of different postures for the attitude of Jesus, leaving it up to the individual worshipper to decide what it means or what is to be hoped for in communing with this personage, who is both mystical and extraordinarily present. Jesus on the cross is no mere symbol that leaves us indifferent. The figure immediately provokes a profound dialogue with what philosopher Paul Ricoeur calls "oneself as another."[16]

Statues of Buddha also have this immediate effect of "emotional contagion," but using different aesthetic means. The head and expression simply suggest serenity: Our gaze and our emotions are subtly directed to and focused on the gestures. Suffering and love are evoked, but internally. The overall message is Contain, master, feel, all in order to share. Everything about Jesus suggests expansiveness, offering, the arms open like the captive before the firing squad in Goya's painting *The Third of May 1808*, or the welcoming gesture you make to a child running to your embrace. Jesus is all-inclusive, joining welcome with suffering and acceptance of powerlessness. He is a bird swooping over the world. Buddha, in contrast, is about contracting, gathering force — like Gandhi, seated and weaving. All internalized tension, concentration, density — Buddha is the universal expressed by the particular, the desire for peace and law that recommends without imposing. The postures of the fingers and the hand carry the essential messages.

Synergies of Bodily Expression of Emotions

Zoologist Thomas Bell and, after him, Charles Darwin described the varieties of motor expression of emotions.[17] But what does this variety conceal? Among others, Jaak Panksepp,[18] a neurologist and specialist on emotions, suggested that, all in all, four fundamental behaviors should be distinguished: seeking; anger, which induces fighting or attack; fear, which produces "freezing"; and finally panic, which can lead to isolation as well as agitation or searching for social contacts. Is this reduction exhaustive? Probably not. Yet we share a large part of our repertoire with primates and even dogs. It is very easy to tell if my poodle, Lolita, is sad or happy, angry or satisfied, fearful or vigilant. According to research, there is a strong resemblance between humans and animals in the perception of postural expressions of emotions and types of defensive behavior.[19]

Today, we understand more about the different networks that contribute to the bodily expression of emotions.[20] We also know that certain areas of the brain contribute to the organization and control of a very precise repertoire of postures linked to aggression and defense.[21] The areas of the cortex that induce these defensive and aggressive postures are very close to the areas that contain the mirror neurons. These neurons activate when you make a gesture or observe the gestures of others. The same regions are involved in the production of behaviors and in perceiving the attitudes of others. A powerful simplifying mechanism is obviously at work here. It is also well established that each of the elements of this repertoire corresponds to a certain sensory organization that does not only measure the movements of the body and its relationships to the world but checks whether *predictions match expectations*. Accordingly, each element of the motor repertoire is also associated with a repertoire of predictions regarding the sensory detector state required for an action to proceed normally. The result is a powerful ability to detect error or novelty, which is useful in reorganizing action in the face of unexpected changes.

This projective and anticipatory biological organization is a remarkable simplification because it allows animals and humans to avoid having to have constant access to a new set of sensorimotor coordinations. The brain is thus free to play its role as predictor, comparator, and decision maker, without having to imagine the best possible solutions in the absence of information. The ability to interrupt an action under way is controlled by the cerebral cortex, which delegates responsibility to lower centers situated in the spinal cord, the brain stem, the basal ganglia, and the cerebellum. In other words, the simplexity of movement is not only expressed in the laws of control that currently keep mathematicians, roboticists, and physiologists busy. A natural and cultural language of gestures is part of the repertoire of techniques that life uses to navigate complex pathways. The major contribution of gesture, in all its richness, is to be immediately accessible both to consciousness, to paraphrase the twentieth-century philosopher Henri-Louis Bergson, and to the unconscious. It is action. Above all, it is acts. As Faust intones in Johann Wolfgang von Goethe's tragic play of the same name, "In the Beginning was the Act."

9. Walking: A Challenge to Complexity

In an astonishing variety of contexts, apparently complex structures or behaviors emerge from systems characterized by very simple rules. These systems are said to be self-organized and their properties are said to be emergent. The grandest example is the universe itself, the full complexity of which emerges from simple rules plus the operation of chance.

—*Murray Gell-Mann*

Walking has played a fundamental role in all animal species throughout evolution. When aquatic life gave way to terrestrial life, a whole set of problems had to be resolved, beginning with integrating the four elements of walking: posture, locomotor rhythm, gaze, and gesture. To the reader who might think that we are about to launch into a question of restricted and basic motor physiology, that is, far removed from cognitive function, I would say that posture is none other than "preparation to act,"[1] and locomotion is not only about making successive steps but also navigating in space. Indeed, some clinicians suggest that the hints of cognitive decline can be seen in the postural and locomotor disturbances of the elderly. And when we are reflecting on a complex problem, what are we doing if not adopting an "attitude," a "point of view," a "posture"?

The Lamprey and the Salamander

The fundamentals of motor coordination for walking have been studied in swimming lamprey.[2] This animal, which has an ancient phylogeny (several million years), propels itself by oscillating movements—special combinations of muscular contractions—produced by "pattern generators" situated in the spinal cord that give different cadences to swimming or locomotor movements (see fig. 10).

It seems that, in the course of phylogenetic history, a basic, hierarchical organization was maintained to produce swimming or locomotion. There were probably important modifications in higher animals in particular during the transition from swimming to terrestrial locomotion or, with the cetaceans, during transition from quadruped locomotion to swimming. The functioning of neural control of the locomotor apparatus in the lamprey illustrates this basic organization very nicely.

Networks in the spinal cord produce the basic locomotor rhythm. The frequency and amplitude of the rhythm are regulated by interactions between excitation and inhibition in the networks but are also influenced by neuromodulators. The menu of the different subtypes of ionic channels that are active in the different neurons is critically important to the functioning of these networks. A simplex property of this system, for example, is its capacity to reorganize its functional anatomy. In fact, the same network can be used to produce different modes of locomotion—the transition from walking to running in humans or from trotting to galloping in horses. Neuromodulatory activities originating in the bulbar centers produce changes in the synaptic connections that are sufficient to change the style of locomotion (walking, running, and so forth). This principle of economy also simplifies neurocomputation.

Coordination between the segments of the body is governed by networks of neurons, also in the spinal cord, that are endowed with

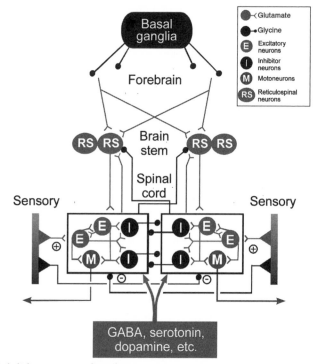

Basal ganglia

Forebrain

Glutamate
Glycine
E Excitatory neurons
I Inhibitor neurons
M Motoneurons
RS Reticulospinal neurons

RS RS Brain stem RS RS

Spinal cord

Sensory   Sensory

E E I   I E
                    E
M I   I M

⊕   ⊕

⊖   ⊖

GABA, serotonin, dopamine, etc.

Adult lamprey swimming

Figure 10. The neural organization of swimming in the lamprey. The study of swimming in the lamprey provides a basic model of the organization of locomotor mechanisms for all vertebrates. This diagram reflects the fundamental principles of this organization. The circles represent the neurons of the motor circuits and their projections. The hierarchical organization of

oscillating properties that synchronize the different segments and regulate timing to ensure harmonious propulsion. The triggering of rhythmic activity is produced by a center situated in the mesencephalic reticular formation. Actually, because the spinal cord oversees the basic organization, the higher centers are "freed" from detailed coordination to attend to global operations. The *basal ganglia*, and in particular the system that links them to the thalamus and the cortex, regulates speed and, no doubt, coordinates posture, intentional movement, and so forth. The *prefrontal cortex* contributes to decisions that also take into account motivational or emotional aspects. The *cerebellum* controls coordination of limbs and posture, learning, and the temporal distribution of motor activity. However, it also is involved in more "cognitive" aspects. It plays a role in the long circuits that connect it to the thalamus and the cortex. Finally, structures such as the *hippocampus* contribute to very global aspects of navigation, that is, planning trajectories, spatial and episodic memory of journeys, and so forth. I propose that this hierarchical organization is a fundamental example of what I call simplexity. If evolution

---

(Figure 10 continued)

swimming is easy to see: the mechanisms for creating rhythms and the coordination of segments in the spinal cord; the supraspinal control of the segments; the mechanisms of selection at the level of the basal ganglia; and, not least, the piloting of all these circuits by the cortical controls of guidance. Also worth mentioning here is the major role of chemical neuromodulation by serotonin, GABA, and so forth. These organizational schemes appear to have been preserved over the course of evolution with modifications enabling, for example, the progression from swimming to walking, running, and jumping. [Reproduced courtesy of S. Grillner and Elsevier. See also S. Grillner, P. Wallén, K. Saitoh, A. Kozlov, and B. Robertson, "Neural Bases of Goal-Directed Locomotion in Vertebrates: An Overview," *Brain Research Review,* 57 (2008): 2–12.]

had tried to have the same network perform such different functions, this network probably would never have been usable. It would have been too complex. Dividing the processors in different modules and coordinating them constitutes an additional apparent difficulty that, in the end, produced a very efficient mechanism that could be repeated throughout evolution.

Another animal, the salamander, puzzled physiologists for a long time. This animal can swim like a lamprey when it is in water; but as soon as it hits the ground, the beach, for example, it begins to walk, using its four feet. These feet produce organized, diagonal movements, and although the rest of the salamander's body undulates, it undulates differently depending on whether it is walking or swimming. How can an animal go from the oscillations of swimming to the precise coordination needed for walking? A recent study showed that this transition is triggered simply by the contact of the salamander's feet on the ground. This contact produces elevated tonic activity in network that is the equivalent of a simple "threshold effect." This simple change switches the network from one mode to the other.[3] What a beautiful example of simplexity!

Planar Covariation

Covariation is a term that has been used for perception by the psychologist James J. Gibson. He proposed that the fact that when we act or move some sensory inputs covary is a powerful aid for the brain to detect invariants in the world. During locomotion, the coordination of limb segments is simplified by a law of "planar covariation." Let us briefly explain the mechanism of coplanar variation: If you move your arm to grasp an object, the measure of the *angle of elevation* of the two segments of your arm and forearm (elevation is the angle that the segment makes with respect to the gravitational vertical) shows that these angles covary. They are linked by very precisely defined phase relationships.[4] Drawing the trajectory of a point whose coordinates are traced in a tridimensional referential and having these angles as coordinates shows that the trajectory of the point is situated on a plane — whence the name "planar covariation" given to this relationship. This is thought to simplify the job of the brain, which does not have to control each segment but can formulate general goals for the movement and let the spinal cord take care of the coordination. The same covariation is found in the angles made by the legs during locomotion[5] — for example, the angles of elevation between the foot, the leg, and the thigh. This law appears in the infant during the first years,[6] at the same time that control of walking is organized. It is therefore genetically determined but is specified during the games and movements the infant undertakes — a nice combination of nature and nurture in the tuning of a simplex mechanism.

Gravity: Ally of Simplexity

The use of gravity by living organisms is perhaps one of the finest examples of simplexity. We know that the vertical force resulting from the acceleration of gravity applied by the earth on our body "glues" us to the earth. Gravity holds fast to the bold animals that took a chance on terra firma. In response, evolution has devised an "antigravitational" control of posture. We are endowed with an exquisitely efficient and unconscious extensor muscle tonus and righting reflexes of vestibular origin; vision gives us a measure of the vertical, and so forth. Recall the times you complained that a package was heavy, or the trouble you had climbing a hill or a stairway. In other words, in life, we devote a certain part of our energy to fighting against gravity and preventing falls, which is the third cause of death after cancer and cardiovascular diseases.

In addition to these mechanisms for fighting against the nefarious effects of gravity, living creatures have found ways of using gravity and the force it exerts on bodies to economize energy. Physicist Murray Gell-Mann has written on this subject: "A wonderful example of the simple underlying principles of nature is the law of gravity, specifically Einstein's general-relativistic theory of gravitation (even though most people regard that theory as anything but simple). . . . Our own species, in at least some respects the most complex that has so far evolved on this planet, has succeeded in discovering a great deal of the underlying simplicity, including the theory of gravitation itself."[7]

But we have learned to use gravity. Gravity is potentially a source of energy. Throughout evolution, biological organisms used available energy—light from the sun, for example. The force and acceleration of gravity, too, were employed in controlling movement. For although it is true that to lift a weight, we have to fight against gravity, which increases weight as a function of mass, we can also make gravity work for us. The builders of pyramids and cathedrals used gravity to lift weight by means of clever devices. Engineers captured mountain streams and rivers to feed electrical power plants by transforming potential energy into kinetic energy and turning the turbines of dams.

It seems that life has also been clever enough to use the available forces provided by gravity. According to a prevailing school of thought in the biomechanics of walking, the control of walking is partly governed by the use of dynamic forces that make it possible to complete muscular activity at certain stages of the locomotor cycle.[8] Roboticists have shown that a headless, spineless robot can be made to walk on a slightly inclined plane and even on a horizontal plane using very little energy. An artificial monkey simulated on a computer can jump from branch to branch using only passive dynamics, provided that it grabs and lets go at just the right time. If it does that, it can ignore gravity. The trick is that the artificial animal must grab or let go of the branches when its speed is zero. The theory that explains this simplification involves concepts of dynamic stability. The idea is that a system can be stable while immobile, but that it can also be dynamically stable. Roboticists have proposed solutions, such as impedance compensation, that allow systems to learn unstable dynamics.[9]

How Not to Fall

As mentioned above, control of posture and equilibrium — a very complex problem — assumes that we fight against the effects of gravity. If not for the sustained activity of our extensor muscles, which resist this force equivalent to an acceleration of 9.81 meters per square second (which is tantamount to going from a full stop to a speed of 10 meters per second in one second), we would be crushed to the ground. Postural control also requires coordination of hundreds of muscles acting in many spatial directions. We say that it has to control hundreds of degrees of freedom.

In humans, regulation must be very fast and very precise, because the support polygon — the surface of the ground that our center of gravity projects to when we fall — is very small. This control is facilitated by many simplex mechanisms. One of them is anticipation. Before lifting your arm, for example, which will take your center of gravity beyond the limits of balance, the brain employs an anticipatory mechanism. Babinski's synergy (named for the French neurologist Joseph Babinski) produces a very slight movement of the body backward, anticipating the displacement of the body forward and preventing a fall.[10] Now, what happens if the ground is slippery or something makes you trip? You have to react very quickly. Faced with this kind of perturbation, your bodily response is often (though not always) reduced to a combination of two movements:[11] One is a movement of the entire body around the ankle, like an inverse pendulum; the other is a movement of the upper body toward the hip. Thus the brain is spared having to construct a three-part, adaptive response. It would never have the time! Instead, it calls on a repertoire of ready-made behaviors to choose from. The sensation that you feel when you are about to fall serves to trigger existing responses that considerably simplify control. This is a further example of simplexity.

Another remarkably simplex solution to the problem of control is what I call transfer of function or *transfer of competence*. Usually, when posture is destabilized and you risk falling, the reflex responses for recovering are activated in the muscles of the legs. However, if a handrail or banister is within reach, the distribution of the reflex activity is completely reorganized: The recovery response is expressed in the muscles of the arms.[12] The brain transforms the arms into legs; it has changed its frame of reference and its fulcrum.[13] In so doing, it obeys a general law: In response to the environmental context or to the properties of the external world, the brain can completely and very rapidly rechannel both motor responses and sensory inputs. Figure 11 illustrates just one of many theories currently being proposed to describe the idea of relative frames of reference. This functional flexibility — known as *vicariance* — is a basic property of the brain.

Figure 11. A variety of frames of reference chosen by the brain to control diverse actions. Three examples of hypothetical reference frames for which the brain may code different types of movements to simplify the computation. (A) To precisely grasp a reference object, R, requires a virtual distance between the thumb and the index finger. The real distance, Q, is compared with this reference distance to produce an adaptive muscular contraction. (B) A movement of a gymnast's legs is produced by a continual displacement of the frame of reference in space. (C) The same applies to movements of the entire body. To reach a cup, the reference system displaces the hand

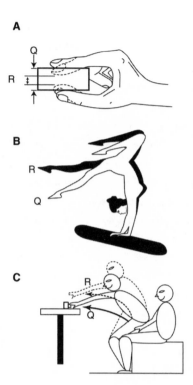

(Figure 11 continued)

toward a reference position in the direction of the cup. The real movements of the hand result from a combination of this movement and principles of minimal interaction between the parts of the body. In a similar task
of "reaching out of reach," we have proposed with Martha Flanders that
the hand movement is also related to gaze. [Adapted from A. G. Feldman,
V. Goussev, A. Sangole, and M. F. Levin, "Threshold Position Control and
the Principle of Minimal Interaction in Motor Actions," *Progress in Brain Research,* 165 (2007): 267–281, by permission of Elsevier.]

Controlling Locomotion from the Head:
A Great Invention

As I said above, gravity provides an ingenious frame of reference for indicating a "vertical," uniformly distributed over the surface of the earth. This "plumb line" is used by the sensors of the inner ear (the otoliths), in cooperation with the sensors for angular acceleration of the head (the semicircular canals). In this way, the brain constantly keeps track of the absolute angle of the head in space, a veritable "inertial guidance system" similar to the ones used onboard airplanes and missiles to stabilize their position during flight. We have shown that in humans, locomotion is not organized from the feet up, which is how most robotic humanoids are currently constructed. Roboticists put the inertial sensors in the trunk. They should be in the head. My theory is that coordination of multiple degrees of freedom of the limbs during locomotion is organized from the head, which is stabilized in rotation and constitutes an inertial guidance platform (see fig. 12). This top-down control acts as a sort of mobile frame of reference and liberates terrestrial animals from the ground.[14] (In mathematics, the great French geometer Élie Cartan also developed a theory of mobile frames of reference.) It is what enables a squirrel to jump from branch to branch with exceptional agility and a bird to fly and to zoom in on its prey with extreme precision, even in very windy conditions. Top-down control also helps surfers, skiers, acrobats, and other athletes to make incredibly complex maneuvers. Without a top-down inertial platform, or locomotor guidance stabilized in rotation, the humanoids currently built by roboticists will never walk as well as humans.

In humans, control of posture and balance is developed in infancy. A baby begins nodding its head and organizing its locomotion through a series of motor and reflex programs that use the ground as reference point and anchor. The least variation in the ground is disastrous for the baby. Between three and five years of age, this little person acquires the ability to run and to jump. He or she becomes "all-terrain." At this stage, a sort of Galilean revolution occurs: The head stabilizes in rotation, and gaze guides position.[15] The infant has freed itself of the earth and becomes a wingless bird. The elegant solution found by evolution to solve this extraordinarily complex problem is not simple at all: Control is shifted from the feet to the head by virtue of a very sophisticated sensory system.

In all likelihood, the vestibular system also contributes to the capacity of all animals to establish coherence among their different senses. This is a fundamental frame of reference. It is not at all surprising that the region of the cerebral cortex that receives vestibular information — the vestibular cortex (see fig. 13), situated at the junction of the parietal and temporal cortices[16] — is just as fundamental for self-coherence.[17] The Montreal neurologist Wilder Penfield confirmed this by electrically stimulating this region in epileptic patients. Based on an astonishing intuition resulting from his observations, Penfield identified this region as being critical to "the awareness of the body scheme and spatial relations."[18]

In other words, the vestibular system is not only a basic frame of reference for stabilizing the head and coordinating movements of the limbs. It is also critical to our perception of space and for constructing our sense of our bodies vis-à-vis the world. Moreover, for birds to evolve from a particular family of dinosaurs (a fact established by paleontologists), the dinosaurs had to have had a vestibular system that freed them from having to rely on the ground as frame of reference. Their vestibular system must have possessed sensors that were geometrically and dynamically very sophisticated.[19]

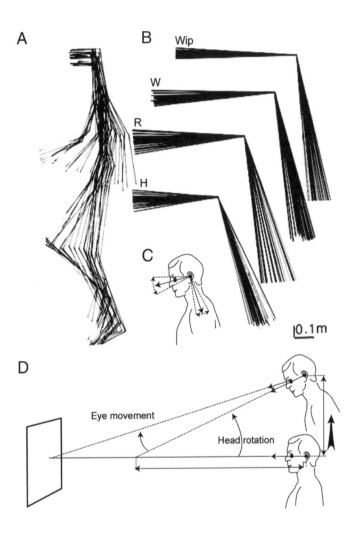

A

B

Wip

W

R

H

C

0.1m

D

Eye movement

Head rotation

Let's Hear It for the Skeleton!

In terms of the body, the twentieth century was paradoxical. It marked the rediscovery of the body's beauty, forgotten since the time of the Greeks under the influence of Judeo-Christian religion, which had made it an object of suffering: The body of Christ is a symbol of pain. Today, we are witnessing a striking return to the body — a veritable orgy. We have the perceiving and perceived body, the dressed or undressed body, the emotional body, the acting body, the body of others.[20] Bodies are groomed, pampered, rejuvenated, shaped and

---

(Opposite) Figure 12. A frame of reference for controlling walking. The head is stabilized in rotation in the sagittal plane (the anteroposterior plane of the body) and serves as a mobile reference frame for the organization of locomotion. This enables the brain to free itself from the reference frame of the ground and avoid having to control the "zero moment point," which requires a ground contact. (A) Computer recording of the movements of body segments during walking. Note the stabilization of the head in rotation. (B) Movements of the segments of the head and trunk during walking in place (Wip), walking (W), running (R), and hopping (H). (C and D) Schematic drawings showing that gaze determines the angle of stabilization of the head. The angular velocity of the head in degrees per second (which constitutes a measure of its rotation) is the same for opened and closed eyes, suggesting that vision is unnecessary for this stabilization, which is mainly controlled by the vestibular system. Stabilization of the image of the visual world on the retina when the eyes are open is governed by the vestibuloocular reflex. This creation of a stabilized platform (the head) allows considerable simplification for multisensory integration and sensorimotor coordination of complex movements, and for top-down cognitive control of locomotion and locomotor trajectories. [Reprinted with kind permission from Springer Science+Business Media, from T. Pozzo, A. Berthoz, and L. Lefort, "Head Stabilization during Various Locomotor Tasks in Humans. I. Normal Subjects," *Experimental Brain Research*, 82 (1990): 97–106.]

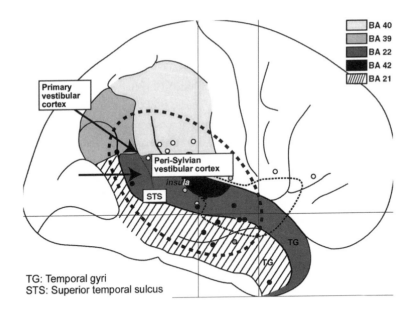

Figure 13. A critical center for self-cohesion and spatial relations of the perceived body: the "vestibular cortex." Proprioceptive, vestibular, and visual multisensory information about the perceived body ("body subject") and its orientation or movements is transmitted to a region situated around the parietotemporal junction. This region is the "vestibular cortex." It is important because it is involved in consciousness (as the neurologist Wilder Penfield predicted), but also in the unified perception of the body (body schema or body image) and the relationships between the body and space. It may also contain a neuronal network implementing an internal model of gravity. In fact, a whole area of the superior temporal sulcus, which we propose naming the "Peri-Sylvian vestibular cortex," is involved in these functions of relating the body to space, of social interactions, of perceiving the actions of others ("mirror system"), and even of empathy. BA indicates the Brodmann areas. [Adapted from P. Kahane, D. Hoffmann, L. Minotti, and A. Berthoz, "Reappraisal of the Human Vestibular Cortex by Cortical Electrical Stimulation Study," *Annals of Neurology*, 54 (2003): 615–624.]

toned, photographed and sculpted, painted, and scrutinized using high-tech imaging. The body has become an object to exhibit and to measure. It is fitted out with prostheses and braces. Even phantom limbs come in for attention. The body's virility is enhanced by replacing vertebrae, teeth, and eyes. Surgery is proliferating. Never have there been so many societies concerned with rehabilitating the body. Never have there been so many "techniques of the body," to use sociologist Marcel Mauss's expression. And never have sports played so major a role or been so much at the center of entertainment.

The body has become an industry. There are thousands of kinds of shoes, clothes, backpacks, hats, equipment for water or mountain activities. The human body is expected to reach the highest summits, the most-distant planets, and the greatest depths. Author Jules Verne would probably be fascinated to know that Captain Nemo now emerges from his submarine and confronts the giant squid armed with electronic weapons and an outfit specialized for extreme exploration, like an astronaut's.

What has happened to the skeleton in this celebration of the body? After all, it is the skeleton that holds the key to simplifying the actions enabled by the body. Muscles could do nothing without this extraordinary structure, which constitutes the fundamental architecture of our physical cohesion, of our relationship to the forces of gravity, and the expression of our repertoire of behaviors. The skeleton is a marker of our *Umwelt*.

To our eyes, a skeleton suggests death. At school it just hangs somewhere in science class, and we are not taught to admire it. But anyone who takes the trouble to go to the National Museum of Natural History in the Jardin des Plantes in Paris will immediately understand what the thousands of skeletons gathered there have to tell us, namely, their *astonishing similarity*. Look at a dinosaur, a bird, a fish, a squirrel, a giraffe. The basic organization is the same. A head, eyes, backbone, respiratory system, feet. Zoologists have spent years emphasizing their diversity. Yet we could equally stress their resemblance.

For the stabilized platform represented by the head to enable coordination of the limbs required a special architecture of the backbone and, in particular, the neck.[21] In most quadrupeds and in birds, the ensemble makes an S of great beauty that creates a horizontal plane for the head and places the vestibular system in a very precise position with respect to gravity (see fig. 14). The horizontal semicircular canal and the utricle are positioned in a horizontal plane where, guided by gaze, they can be stabilized at a specific angle. This neck architecture is accompanied by a very elegant distribution of the muscles. Look at a pigeon or magpie walking. The head bobs back and forth (head nystagmus). When the body moves forward, the head retreats. This stabilizes the image of the world on the retina because the eyes are lateral. The S geometry of the neck also contributes to this remarkable simplification—yet another example of simplexity.

(Opposite) Figure 14. The marvelous simplex architecture of the neck bones. During evolution, the architecture of the skeleton of the neck was modified to maintain the horizontal position of the skull during normal walking and the stabilized head as described in figure 12. This striking and complex architecture permits vestibular stabilization of the head in the horizontal plane; for orienting gaze movements, there is actually a network specialized in horizontal movements located in the brain stem. The geometry of the bones of the neck makes it possible to dissociate horizontal, vertical, and twisting movements, which introduces complexity. But it also makes control considerably more refined and efficient: It is simplex. [Adapted with kind permission from Springer Science+Business Media from P. P. Vidal, W. Graf, and A. Berthoz, "The Orientation of the Cervical Vertebral Column in Unrestrained Awake Animals. I. Resting Position," *Experimental Brain Research*, 61 (1986): 549–559.]

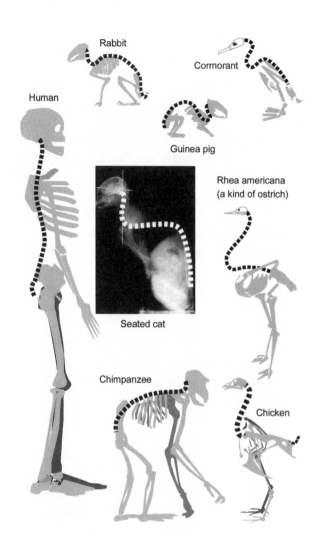

Rabbit

Cormorant

Human

Guinea pig

Rhea americana
(a kind of ostrich)

Seated cat

Chimpanzee

Chicken

Do the Same Laws Govern Movement of the Arms and Walking?

Walking is not simple. You do not just have to keep your balance and make steps. You have to go somewhere, following a path. And you have to use your body to produce a trajectory like the one I described above using a hand to draw. I propose the following simplex idea: *The laws that underlie movement of the hands are very similar to those that govern the production of locomotor trajectories.*

Let us take the example of a hand movement, a natural gesture. If you wish to pick up a glass sitting on a table, there are an infinite number of possible paths to it (at least in principle). The same is true if you wish to cross an empty room and exit through a door. In both cases, however, it turns out that the trajectories are strictly defined. Evidently, precise laws determine this stereotyping. Psychologist Roger Shepard suggested early on that we take "the simplest route" to execute a displacement.

One of the constraints on the trajectory is biomechanical: Our skeleton does not allow certain movements of the arms, and when we walk, we cannot easily move sideways, like crabs. In general, we walk in a straight line, putting one foot in front of the other. Rarely, we might make a step to the side. Roboticists say that we are partly "nonholonomic." A car is completely nonholonomic because it cannot move sideways at all. But some of the laws of natural movement do not only originate with biomechanics. They originate with the very functioning of our brain. Hand movements, like locomotor trajectories, result from neural processes that optimize values related to kinematics—for example, minimizing jerk—or that minimize the energy required.

The focus of current research is to discover these optimization laws. Recently, we suggested that the locomotor trajectories that we adopt very closely resemble segments of spirals.[22] Other studies indi-

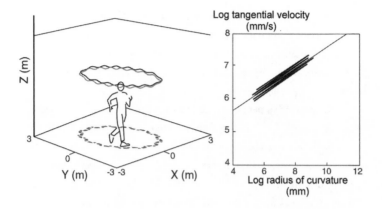

Figure 15. The same laws govern movements of the arm and production of locomotor trajectories. The one-third power law, which connects tangential velocity and the curvature of a trajectory, also holds for locomotion. (Left) A person following an elliptical path as well as the movements associated with his head, measured by an optical system that is linked to a computer. (Right) The linear relationship between tangential velocity of the head along a trajectory and the curvature of displacement of the head in the horizontal plane during walking along this elliptical path. [Adapted from S. Vieilledent, S. Dalbera, Y. Kerlirzin, and A. Berthoz, "Velocity and Curvature in Human Locomotion along Complex Curved Paths," *Neuroscience Letters*, 305 (2001): 65–69, by permission of Elsevier.]

cate that for the hand, the forms are partial parabolas,[23] and that these parabolas result from the brain's use of several geometries.[24] Recall that in drawing or writing, the tangential velocity and the curve of the trajectory of the finger are linked to a very simple law called the one-third power law,[25] according to which we go more slowly when the curve is large. This law of natural movement is important because it also governs our perception of the movement of others.[26] If the movement is in space and the torsion of the limbs intervenes, of course we need to complicate the mathematical formula a little. But the relationship remains simple.[27] This law was originally formulated for hand movements. We have shown that it also applies to walking trajectories when they are simple, for example, an ellipse followed on the ground (see fig. 15).[28]

However, if someone is asked to follow more complex trajectories, a figure 8, for example, little variations of the coefficient that is no longer one-third become obvious.[29] Often, in science, small variations in numerical values in laws mean that the phenomenon is still being described more or less well but that the general theory that forms the basis for the method of observation is not the right one. That is why we have hypothesized that these little variations in coefficient indicate a major revelation, namely, that the brain does not rely only on Euclidean geometry.[30] In fact, it seems that the brain combines Euclidean, affine, and pseudo-affine geometries and varies their respective contributions depending upon speed, curvature, and so forth. This apparent complexity is probably extremely useful in the control of locomotor trajectory: It obeys the principles of simplexity.

Part III. THE REALM OF THOUGHT

10. Simplex Space

Extensity, *discernible in each and every sensation, though more developed in some than in others,* is the original sensation of space, out of which all the exact knowledge about space that we afterwards come to have is woven by processes of discrimination, association, and selection.

— *William James*

Let us now quickly review the neural basis of spatial processing in the brain. The goal here is not to present a course on physiology but to show how the spatialization of perception, action, memory, and decision making reduces complexity, sometimes by way of detours that, in turn, engender simplexity.

The Brain Is a Geometrical Machine

This theory has been revisited several times since neuroscientist and, later, Nobel Laureate Santiago Ramón y Cajal revealed the remarkable diversity of neuronal morphology. It seems obvious when we recall that the body, or even the outside world, is represented in the brain by neural maps organized by "topies," which means that the neurons that receive information from the body or the world are distributed in the brain centers on maps that replicate the layout of the body or the world. For example, in the visual cortex and

Figure 16. *The Devil in Paris; Paris and Parisiens: Customs and Mores, Caricatures and Profiles of the Inhabitants of Paris.* The devil of Paris is a surveylike, or "allocentric," representation of the city, as opposed to the route-based, "egocentric," representation he would construct walking about Paris without a map. [Illustration by P. Gavarni, published by J. Hetzel (1845).]

the superior colliculus, the neurons that receive information from the visual world are laid out according to "retinotopic maps." The same is true for the sensory cortex, or the cerebellum. This is also true for some brain areas involved in commanding movement. An example is the "homunculus" in the motor cortex (somatotopy). We speak of a homunculus because the pyramidal neurons of the motor cortex that direct limb movements are arranged according to a map of the body.

To understand the notion of space, it is important to realize that *spatialization is a fundamental property of life.* As neurobiologist Alain Prochiantz explains in the *Les Anatomies de la pensée* (The Anatomies of Thought),[1] developmental genes are arranged on the chromosomes collinearly with the spatial organization of the body: a genetic homunculus, so to speak. The constraints of time and space imposed on the expression of "homeotic" genes (a class of genes that encodes positional information, allowing the appropriate spatiotemporal decisions to be made by cells and tissues) are reflected in the genome. Moreover, the genetic organization of the body includes a hierarchy of maps of various shapes that coincide from the spinal cord to the cortex. In other words, despite the diversity of the arrangements of these maps, the same basic organization is found at different levels of the brain, which must simplify the registration of the maps.

The genetic homunculus is not unidimensional but multidimensional: As Prochiantz puts it, "spatial information is primordial." The differentiation of the anterior brain responded to the need to link sensorimotor order reflexes (whose encoding is spatial) to other sensory modalities, such as sight, hearing, and smell. Consequently, it is also possible to find deformed homunculi that are related to the genetic homunculus represented in the genome, in the case of the vertebrae, by the four HOX complexes (see chapter 2).

Space: A Common Way to Encode Perception

The brain uses space to encode sensory inputs, for example, to extract from the immense and subtle variety of sounds those that are pertinent and important in the animal and human kingdom. This enables each individual to recognize its congeners. A penguin, for example, can distinguish the sound of its own infant's cries among thousands of others. In the cochlea, receptor neurons that effect the mechano-neural transmission of sound vibrations detected by the ciliary hair cells are placed in a specific order on the scala vestibuli, a marvelous stairway of sound. This is how, in this case, space helps to encode the vibration frequency. This spatial ordering of sensory stimuli is a fantastic invention of evolution. It occurs notably in encoding odors through the olfactory glomeruli, the peripheral neural center in which odors are processed. It reinforces the idea that the brain is a machine for which geometry plays a major role. The same is true of the vestibular sensors, whose complex geometry is adapted to three-dimensional perception of head movements. In my opinion, this spatialization serves to simplify neurocomputation. It gives the brain a universal "code" that applies from the very first levels of sensory detection, whether visual, auditory, olfactory, tactile, or vestibular. And this universal code or "language" clearly simplifies central processing. In a way, it is what people discovered long before computers when they used abacuses to count. It is also basically what writing does in spatializing an idea or a material fact. Finally, it is the purpose of symbols, or of art, which is always spatial, and music, since the notes are arranged on a staff. The composer Pascal Dusapin has shown most impressively in his lectures at the Collège de France how he uses space to create a symphony and an opera.[2] A page from a score of a symphony is a simplex space.

The Problem of Multiple Geometries

It is hardly surprising that we are often disoriented, overcome by vertigo, or that we forget our keys or get lost. The brain must integrate a great deal of information provided by the senses and select from it based on our goals, our pleasures, and our fears. All sensory detectors operate within different frames of reference. They "encode space" according to a variety of geometries. For example, the tactile sense encodes variations in pressure or the shifts in contact on the supple, changing surface of skin. Muscle receptors (neuromuscular bundles) are clueless about the surface of the skin. They only understand the stretch of muscular fibers. The receptors in the joints measure the angles between the limbs, but they are unaware of the muscle stretch. The vestibular receptors of the inner ear measure head movements in the three perpendicular planes of the canals and the otoliths that form a Euclidean frame of reference. (And who knows? Perhaps this reference frame is the origin of Euclidean geometry.)

Vision encodes an image of the world on the spherical retina. We call this spatial encoding "retinotopy." It is found in the primary visual cortex (V1). For example, in the primary visual cortex of higher mammals, the neurons are arranged according to a directional map (this is well established, beginning with the work of neuroscientists David H. Hubel and Torsten Wiesel), but within this map, the neurons are also arranged according to a pinwheel geometry, such that the directions radiate from a center.[3] In this pattern, the neurons are highly sensitive and selective with respect to orientation, even at the center. The pinwheels not only simplify neurocomputation; they help V1 to process geometric structures. Indeed, for a network of neurons to be able to identify geometric structures (for example, virtual contours), the network must have a very precise functional architecture. It is believed that the pinwheels implement neurally what is known in differential geometry as the "contact structure" of the curves of a plane, which enables V1 to detect the contours.[4] The organization of visual pathways enables signals to be processed in different frames of reference, depending on the task assigned to each module of the brain's visual analyzers.[5] These maps are sometimes altered to fill particular functions. For example, in the visual cortex, enlargement of areas dedicated to the fovea increases visual analytical capability without having to increase the number of cells tenfold.[6]

It is hard not to be impressed by the diversity of neuronal geometries, or the many different arrangements of neurons in "columns" in the visual cortex and inferotemporal cortex, where the shapes of objects are processed, and in groupings and specialized domains in the thalamus and the basal ganglia. Where the configuration does not follow a visibly "geometric" organization, rules of functional proximity prevail, as in the precerebellar nucleus of the brain, called the olivary bodies, where the electrical connections between neurons require nearness. Finally, the organization of axonal branching subtends the geometry of movements and simplifies neurocomputation. In the cerebellum, the geometrical disposition of the Purkinje cells resembles a line of pear trees in a well-tended orchard. The network of connections for inputs to the "mossy fibers" on the Purkinje cells is another spectacular example of the brain's geometric organization. The list could be extended to include the types of arrangements in the basal ganglia and the substantia nigra, where precise mapping enables control of structures such as the colliculus. The challenge here is to understand how these different geometries solve different problems and also how the brain is able to build a coherent perception of space from this diversity.

The Curious Geometry of the Colliculus

A great many animals do not have a visual cortex as developed as that of primates and humans. In all animals, however, the image of the external world is projected onto another map, a sort of internal retina called the tectum, whose name is also, as we saw in previous chapters, the superior colliculus. It is in this structure that the animal constructs a response to the appearance of a moving visual target at the periphery of its visual field. The superior colliculus enables the animal to rapidly orient the eyes, the head, and the body itself toward the object that is approaching and, in an anticipatory way, to capture or avoid it. This map loses the simplicity of the spherical geometry of retinal projection; it is not spherical anymore.

In the superior colliculus, the visual world is represented (see fig. 17) according to complex logarithmic mapping. Why? Perhaps, as my co-workers and I have shown,[7] this deformation makes it possible to optimize sensorimotor transformations. This special arrangement surprised us, and we showed that this map results in a combination of properties that facilitate control of saccades. The transformation of the geometry of a central representation to facilitate execution of a very rapid movement is a good example of simplexity. At the cost of a complication—transforming the geometry—the brain resolves a difficult problem in a very elegant way: It matches a sensory space with a motor command in a complex muscle space. It is a simplex transformation because it solves the problem very efficiently by introducing a modification. Moreover, the arrangement of the neurons in the collicular map varies with the animal species, probably to accommodate the repertoires of behaviors of each animal and its *Umwelt*. Constructing a unique and coherent perception of the body is thus a remarkable property. We still do not know precisely how what roboticists call "sensor fusion" takes place. Most likely through simplex mechanisms.

Near and Distant Spaces: A Matter of Interlocking

The problem of matching spaces is not merely a sensory issue. The brain also has a specialized network for our movements in space. Look around you: Whether you are in the city or the country, the environment is incredibly complex, and you have to process very different problems depending on whether you are tying your shoes, picking up an object, or beyond bodily space, stretching to retrieve something, plucking fruit from a tree, walking to a nearby location, or traveling several kilometers to get home.

Faced with the multiplicity of spaces in life, evolution resolved the complexity by deconstructing the problem. The brain has different ways of processing action that takes place in the space near where you are sitting or that you might be planning farther away. Simply pointing to an object at 70 centimeters (within reaching space) or at 2 meters (within room distance) activates different neural networks in the brain.[8] Similarly, the brain activates different networks depending on whether you are thinking about walking from your chair to the door or, in the city, from the house to the office. Why this modularity? Does each space require a different type of processing?

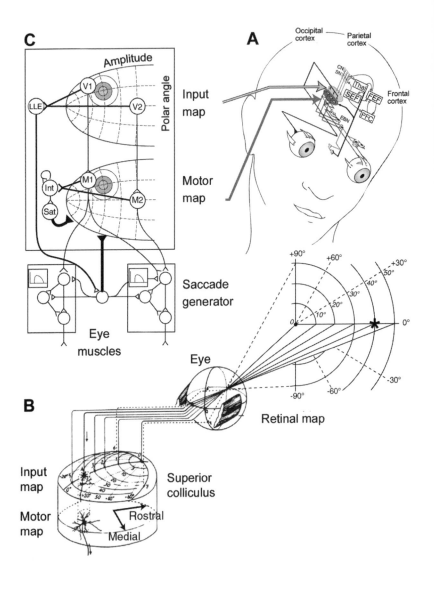

C

Amplitude

V1
V2
LLE
Input map

Polar angle

Int
M1
M2
Sat
Motor map

Saccade generator

Eye muscles

A

Occipital cortex
Parietal cortex
Frontal cortex

CN
SN
Thal
SEF
FEF
PFC
EBN

Eye

+90°
+60°
+30°
50°
40°
30°
20°
10°
0°
-30°
-60°
-90°

Retinal map

B

Input map

Motor map

Superior colliculus

Rostral

Medial

Perhaps the geometries in which networks work are not the same — Euclidean for some, affine for others, and so forth?[9] Computational neuroscientist Tamar Flash and her collaborators have developed the theory that arm movements are controlled by transformations of the affine plane that preserve the area (that is, equi-affine geometry).[10] She and Frank Pollick, a psychologist, independently proposed that the two-thirds law in motion production is related to affine invariants. Using brain activity recordings from functional magnetic resonance imaging (fMRI), she also confirmed that different brain networks are activated by motion perception corresponding to Euclidian or affine geometries. Perhaps the three geometries are not present in all these structures? Time will tell.

This multiple encoding of space doubtlessly simplifies neurocomputation, but it comes at a cost: It requires that all the different areas be synchronized to enable them to be successively or simulta-

---

(Opposite) Figure 17. The curious geometry of the superior colliculus. (A) The superior colliculus is composed of several layers (superficial, intermediate, and deep) that constitute input sensory and output motor maps of visual space. These maps represent the retinal map centrally. (B) The retina codes visual information according to a spherical geometry, but in superficial (and deep) layers of the colliculus, the "neural images" on the retina are arranged according to a complex logarithmic geometry. [Redrawn with permission from A. Grantyn.] (C) A model has been made of the correspondence between the geometry of the colliculus and the motor apparatus of gaze control between vision and the muscles of the eye. "Amplitude" and "polar angle" indicate that in both the superficial and deep layers of the colliculus, visual signals can be coded according to a polar coordinate system. [Adapted with kind permission from Springer Science+Business Media from N. Tabareau, D. Bennequin, A. Berthoz, J.-J. Slotine, and B. Girard, "Geometry of the Superior Colliculus Mapping and Efficient Oculomotor Computation," *Biological Cybernetics*, 97 (2007): 279–292.]

neously activated. When you are driving, you sometimes keep an eye both on the GPS and the road. I think that in many psychiatric and neurological diseases, such as childhood motor-skills disorders, it is these different frames of reference that are disturbed and that produce the apparent "motor" symptoms. For instance, we are working with developmental neuroscientist Giovanni Cioni in Pisa to show that children with cerebral palsy who have difficulty walking have not only motor disorders that prevent them from walking but have also spatial cognition disorders and difficulty in manipulating spatial reference frames.

Although multiple encoding of space introduces complexity, it is also a solution for rapidly and simply solving problems within each space—personal, extrapersonal, distant—and this modularity is characteristic of simplexity. Where it exists, neurocomputation is simplified. My own view is that geometry is a tool of simplexity. This question has recently been the subject of discussion,[11] although no proof is yet available. Nonetheless, I am convinced that the future will show that the use of geometry, and thus of space, to organize neural activity leads to remarkable simplifications regarding brain processing, flexibility, and adaptability. One of the most dramatic achievements of evolution is to have transformed the very rigid spatial organization of the nervous system, determined by the genes, into an exceptional tool that allows us to mentally manipulate space and, especially, to create concepts. The formula "We think with our bodies" does not capture what I mean. The brain "creates worlds" based on the body in action in the world by virtue of the flexibility and multiplicity of its mechanisms for manipulating space.

The diversity of maps, frames of reference, and modes of encoding is fundamental to our capacity for mathematics, reasoning, and thinking. One of the great advances of evolution is the use of these specialized processors—this extraordinary variety of spatial encoding—and the ability to organize the external and internal world as a veritable society of specialists by combining competences and languages, giving unparalleled freedom to the brain's workings. The extreme specialization of the representations of space gives rise to rigorous analytical grids that contribute to what I call simplexity.

Changing Point of View

Changing perspective is a basic mental faculty of the brain. Imagine that you are in a city you do not know and you are leaving your hotel to go to a museum. You will remember your path by some cognitive strategy (the sequence of streets and locations you passed through or by using a map). Now, after the visit, you want to come back to your hotel. You must change both the representation of your path and "points of view." To get back, you will have to invert the perspective you formed of landmarks and places. This change of perspective requires a *mental rotation* of the scene. The existence of a mental process specific to changes in point of view has been confirmed by research in experimental psychology. Here, too, there seem to be a number of possible cognitive strategies for changing point of view. Suppose, for example, that you are sitting in your armchair, observing an object in the room, another chair, say. Now close your eyes, get up, and make a few steps around the room. At the end of your little walk, still with your eyes shut, try and remember the position of the chair, for example, with respect to the window. Is there only one way to do this exercise? Actually, no. You have a choice between two strategies: remembering the chair throughout the perambulation—a strategy of continuous updating—or forgetting the chair during your walk so you can concentrate on your movements, and at the end, doing a mental recalculation. The brain has several solutions for resolving the same problem. Consequently, it can use all the modules mentioned earlier, in particular, taking into account the context of the task or of one's own state. This flexibility and redundancy are some of the basic properties of simplexity.

I think that the same mechanism is involved in understanding others, and it is why I have proposed a spatial theory of empathy,[12] based on the human capacity to manipulate point of view. Empathy is critical to social relations and divining the intentions of others. It is also essential to rational thinking because it enables us to examine a variety of facts and arguments. This mental operation assumes that we make a sort of mental rotation on ourselves with respect to the environment or an object in it, all the while maintaining an overall perspective.

Involvement of the hippocampic formation in changing perspective has been shown in a study of patients undergoing hippocampic resection.[13] The paradigm consisted in showing a subject a virtual room in which a virtual clown was standing. Virtual reality specialists call these characters "avatars." The room also contained a lamp positioned on the floor. The direction of the avatar's gaze indicated the new point of view from which the room would be presented in the succeeding image. This anticipation improved the performance of healthy subjects. However, patients with damage to the hippocampus could not detect the different between the new and old points of view. Anticipation of the new point of view was imprecise, and patients with damage to the left side of the hippocampus had only a vague idea of what the avatar was "seeing." To my mind, this ability to change spatial perspectives is an essential component of the differences between the sexes.

Do Men and Women Employ the
Same Simplex Mechanisms?

The impact of psychiatric disturbances such as depression, spatial anxiety, agoraphobia, or autism is very different depending on the sex. Today, it is acknowledged that these differences are linked not only to variations in performance but also in mental processes, in many cognitive tasks involving space.[14] For example, men are better than women at tasks of mental rotation, that is, requiring a change in point of view. They also do a better job at remembering maps and in switching from the virtual to the real. Women are better at identifying objects that have moved, and are faster in tasks requiring discrimination, comparison, and verbal mediation. They can do several things at a time. They also perform better in memorizing important shapes especially if the shapes have a name.

Among other things, these differences probably have a hormonal basis. The menstrual cycle clearly influences navigational performance due to varying levels of estrogen, and this modulation is also found in men together with the fluctuations of testosterone. Nonetheless, hormonal factors are only partly responsible, and although puberty is an important period for demarcating spatial ability, some differences appear before and even very early in infancy. The parietal cortex may be at the origin of these differences between the sexes. Its two hemispheres are asymmetrical, and it is possible that these asymmetries are regulated by androgens, which are known to induce an increase in the size of the right cortex in males. Whatever the ultimate explanation, one fact is indisputable: The brains of men and women present anatomical and functional differences that should be taken into account in understanding how they learn and work.

Multiple Frames of Reference and Cognitive Strategies

To remember the way from home to work, you can memorize the route based on the movements you have made and the places you associate with them ("turn left at the bakery"): This is what I call an egocentric and topokinesthetic route strategy (see chapter 2, fig. 4). Another way to do it would be to recall the route using a mental map. This map strategy is allocentric (you don't use yourself as a reference) and topographic (it is akin to a map or a drawing). Each of these two strategies mobilizes distinct regions of the brain. A parietofrontal (mainly the right) cortical network is specialized in egocentric perception of the relationships of the body and space, whereas another network, which includes some of the first but also the temporal lobe, is active in allocentric perception of space.[15] The retrosplenial cortex is mobilized during changes in point of view. This part of the brain is interesting because it belongs to a very fascinating network of neurons called "head direction cells," which code the direction of the head in space even in the dark. They allow you to find the light in the bedroom at night! The retrosplenial cortex is believed to be involved in transforming egocentric encoding into allocentric encoding. It also is involved in processing emotions: an interesting "carrefour" between the emotional brain and the cognitive brain.

Now, let us do an experiment. Choose two objects on a table or in the room where you are reading. First question: Which of the two objects is closer to your body? Answering requires you to perform an egocentric task, because you have to figure out where the object is relative to your body from your perspective. Second question: Which of the two objects is closer to the front wall of the building in which you are seated? This time, answering requires that you carry out an allocentric task. You have to reconstruct the entire building exclusive of yourself. All that matters is the relationship between the object and the wall. Imaging recordings of brain activity show that different networks are involved in each of these tasks.[16] Here, again, we encounter the principle of *modularity of function*.

Another example will illustrate further the fact that our brain uses different networks for dealing with different cognitive spatial strategies. In the following experiment, subjects placed in a scanner (functional magnetic resonance imager) are shown a virtual palace. The courtyard of the palace contains three objects: two trash cans and a ball. Next, the subjects are asked two questions relating to the distance of the objects with respect to the subjects: First, Which trash can is closest? This is an egocentric task because it requires the subject to situate the trash can relative to their own body. Second, Which trash can is the closest to the largest wall of the palace? This time, the task is allocentric. It requires the subject to imagine the totality of the palace—a little like me asking you to think about whether this page is parallel to a street or an avenue of the city where you are. Figure 18 shows the areas of the brain activated in this experiment.

In the figure, it is obvious that the two tasks, which correspond to mental operations in different spatial frames of reference, activate different networks in the brain. The egocentric task is carried out by a parietofrontal network, which is damaged in disorders such as spatial negligence. Patients with lesions in this network on the right side of the brain cannot imagine the right half of the visual world. The allocentric task calls into play a parietotemporal network as well as an important area called the retrosplenial cortex (mentioned above). The brain deconstructs problems linked to space and processes each category of problems using specialized modules. Thus, there is no absolute or relative space. There are several "action spaces" specific to each species, which probably also are processed differently. We do not yet fully understand the laws governing these processes or how they work.

B Egocentric Lateral view

A

C Allocentric Medial view

Figure 18. Two different brain networks are activated according to two cognitive strategies for representation of space (activity measured by magnetic resonance imaging). (A) Courtyard of a virtual palace that contains three objects: two trash cans and a ball. Prior to brain recording, a subject is familiarized with the global architecture of the palace by navigating through the courtyard in virtual reality. The subject is then asked to complete three tasks: a visual control task (Which trash can has fallen over?); an egocentric distance-evaluation task (Which trash can is the closest to you?); and an allocentric task requiring mental manipulation of the environment (Which trash can is the closest to the longest wall of the palace?). (B) Schematic representation (lateral view of the cerebral cortex) of activation related to the egocentric task in a parietofrontal network. (C) Schematic representation (medial view of the cerebral cortex) of activation for the allocentric task. Shown is a portion of the parietotemporal network with activation of the retrosplenial cortex and the parahippocampus. [Adapted from G. Commiteri, G. Galati, A. L. Paradis, L. Pizzamiglio, A. Berthoz, and D. Le Bihan, "Reference Frames for Spatial Cognition: Different Brain Areas Are Involved in View-, Object-, and Landmark-Centered Judgments about Object Location," *Journal of Cognitive Neuroscience*, 16 (2004): 1517–1535.]

The Hippocampus: A Cognitive Map?

Contrary to what one of my visiting students believed, "hippocampus" does not only signify a sea creature. It is also a brain structure that processes space and memory of the routes we take, called "navigation" — in other words, the action of moving about the world, the city, or the country. The hippocampus possesses properties that simplify neurocomputation for navigational memory; it contains neurons that are active when we are in a very precise place. They intervene to establish an "allocentric cognitive map" of our displacements, playing the role of topographic memory (whence the name "place cells"). We have shown that as rats develop, the cells' precise encoding of places becomes more refined.[17] We infer that a similar increase in the precision of localization occurs in infants. But the hippocampus does not only encode places. It also contains "edge cells," which are active when an animal is moving around the perimeter of an arena, for example. Several functions have been attributed to the hippocampus. It is beyond the scope of this book to describe them in detail. Let us just say that the hippocampus is an essential structure for processing space and that it is also appears to play roles as a "cognitive map" and in detecting novelty and conflicts (thus an anxiety trigger). The hippocampus is also linked to "episodic memory," that is, memory of events that occur in going to a specified place. Finally, as we shall discuss below, the left hippocampus, perhaps in association with the cerebellum, is specifically involved in "topokinesthetic memory," that is, memory of the succession of movements that we make along a route (going straight, turning right), associated with this or that event or this or that visual landmark.

Global and Local: Maps and Directions

Try to remember the route from your house to work. On the way, you recognize objects and experience events. Do you remember the order in which you see things or the precise succession of events? How does the brain manage both to remember and to process two types of information: egocentric information, which establishes the sequence of events, and allocentric information, which provides an overall view of the route? How on earth does it deal with this much complexity? While driving, you have a GPS to distinguish the two functions. But the brain?

We have shown above that different networks are doing the job. But, in addition, the two brains, left and right, are not contributing to the same processes. The solution devised by evolution is lateralization. The left cerebral cortex seems to be involved in egocentrically processing, in the first person, the details of a scene. It takes care of classifying, sequential ordering, and narrating directions ("I turn right to the bakery, then I go straight, then I go left at the fountain"). The right cortex appears to be involved in processing allocentric, global, and measurement aspects of space. This division of labor may be due to differently sized "receptor fields" in the two cortices. In reality, the different cortical stations involved in processing spatial data in cooperation with the hippocampus are organized in a network. In other words, you process successive events with your left brain and global aspects of the route with your right brain.

We find this dichotomy, which is obviously very schematic, to be especially interesting because it is simplex. The two brains must work together and exchange information through the corpus callosum, a veritable information highway between the two cortices. Nonetheless, I believe that this mode of functioning is more efficient and functional, even if it introduces an additional difficulty: The exchange of information assumes a shared encoding of information or, at least, compatible encodings. Neurologically speaking, deficits induced by the absence of transfer of information between the two brains are well known.

To prove this lateralization of function, we carried out a navigation experiment in a virtual city with a team of neurologists from the Salpêtrière Hospital in Paris (see fig. 19). We hypothesized that in humans, a lateralization produced a right hippocampus involved (like the right brain) in global, allocentric aspects of spatial encoding, and a left hippocampus concerned especially with encoding sequential, egocentric aspects of routes and the memory of events occurring along the way.[18] This would also implicate the left hippocampus in episodic memory and memory of a train of events en route, corresponding to what we observe in language when we orally relate a trip. We described the stages in sequence (going straight, turning right, and so forth), encounters (a bakery, a friend), and landmarks (a fountain). We reported a third strategy that is also sequential and that relies on the continuous control of navigation and body movements, all the while using environmental clues. The cerebellum may come into play in this process.[19]

Space Is Encoded as a Grid

A recent discovery revolutionized how we think about the way places are encoded. "Grid cells" in the entorhinal cortex have a special property. Unlike the neurons of the hippocampus, they are activated not when a person stays in one place but when a person is in several places in a room. In the case of a rat taking a walk, the places where the cells are activated form a very regular "grid." The active areas are distributed in a network of places very regularly disposed, as if they were distributed in a net of equilateral triangles roughly 30 centimeters on a side (see fig. 20). These neurons detect the geometry

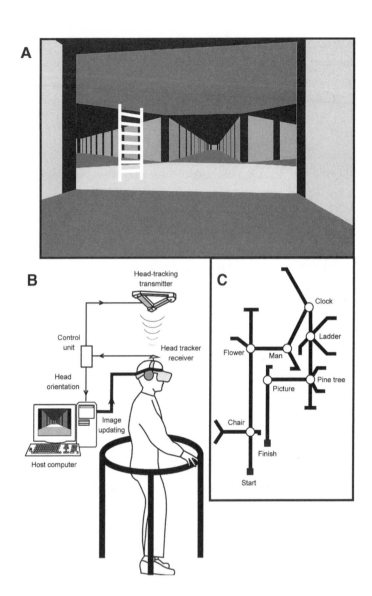

(Opposite) Figure 19. A virtual reality setup for studying cerebral lateralization during navigation. This figure illustrates an experimental design for establishing lateralization of hippocampal functions in remembering trajectories. The exercise consists in traversing a very simplified virtual town — shown in (A) — that contains only streets and a landmark (object) at each intersection. The translations are dictated by the computer, and the rotations are performed by the subject himself by turning his rotating chair — shown in (B). An ultrasound sensor measures the rotations and updates the virtual scene. Once the subject has traveled through the virtual town, he is asked to describe the objects he saw at each intersection (episodic memory). (C) The map of the virtual town. Subjects with damage to the left hippocampus exhibit memory deficits regarding the order of the objects encountered and, especially, the association between the object and rotation of the body (turning right or left) — typically a hippocampal function. Based on these results, with S. Lambrey we hypothesized and showed that the left hippocampus is specialized in sequential memory of events along a trajectory (episodic memory), whereas the right hippocampus is more involved in the allocentric, cartographic (overall) aspect of remembering the route. This was further demonstrated in recent work with functional brain imaging (Iglói et al., "Lateralized Human Hippocampal Activity Predicts Navigation Based on Sequence or Place Memory," *Proceedings of the National Academy of Sciences of the United States of America*, 107 (2010): 14466–14471). [Adapted from S. Lambrey, M. A. Amorim, S. Samson, N. Noulhiane, D. Hasboun, S. Dupont, M. Baulac, and A. Berthoz, "Distinct Visual Perspective-Taking Strategies Involve the Left and Right Medial Temporal Lobe Structures Differently," *Brain*, 131 (2008): 523–534, by permission of Oxford University Press.]

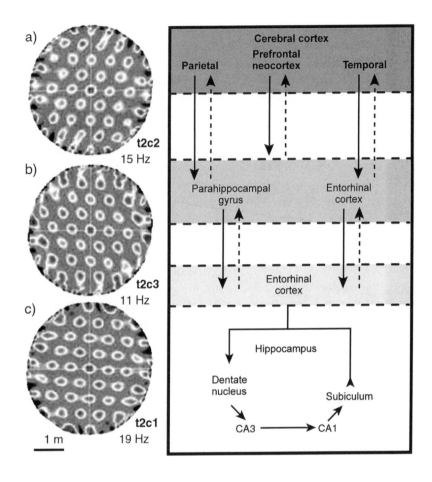

a) t2c2 15 Hz

b) t2c3 11 Hz

c) t2c1 19 Hz

1 m

Cerebral cortex

Parietal — **Prefrontal neocortex** — Temporal

Parahippocampal gyrus — Entorhinal cortex

Entorhinal cortex

Hippocampus

Dentate nucleus

Subiculum

CA3 → CA1

of a space according to very specific rules,[20] and their arrangement in the cortex is mostly likely important to this function.

I propose that all the structures of the brain where neurons operate according to specific geometries are intrinsically organized to enable processing of specific aspects of signals coming from the exterior world or from inside the brain. I even wonder whether the pleasure we take in contemplating floor tiles does not reflect this grid encoding!

Sleep Aids Memorization

One great mystery remains to be elucidated: How are the signals mapped out in all these specialized structures put back together? Another way of expressing this process is "linking." Modern theories suggest that synchronized oscillations have a major role in solving this problem: Neurons simultaneously activated in different areas of the brain are synchronized by other neural activity resulting from in-phase oscillations. We have already seen the importance of oscillations in constructing perception. One of the basic properties of neurons is that they are the seat of oscillatory activities. The brain can be thought of as consisting of coupled oscillators. In his remarkable book *Rhythms of the Brain*, neuroscientist György Buzsáki suggests

---

(Opposite) Figure 20. Grid encoding of space in the entorhinal cortex. This figure shows the remarkable regularity of the activity of so-called *grid cells*, which are neurons in the entorhinal cortex. The first column shows the activity in three neurons of the entorhinal cortex of a rat moving around a circular ring roughly 2 meters in diameter. The places where the neurons are active are shown by the small circles that are surrounded by white. These "grid cell" neurons activate according to a very regular geometry (at intervals of around 30 centimeters), which constitutes a genuine grid specific to each neuron. [Adapted from T. Hafting, M. Fyhn, S. Molden, M. Moser, and E. Moser, "Microstructure of a Spatial Map in the Entorhinal Cortex," *Nature*, 436 (2005): 801–806, by permission from Macmillan Publishers Ltd.] At right, a schematic diagram showing the strategic position of the entorhinal cortex in the circuits linking the areas of information processing (parietal, prefrontal cortex, and temporal) and the areas of the hippocampal formation (here represented separately).

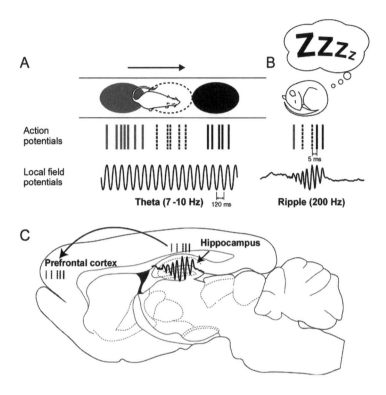

A

Action potentials

Local field potentials

Theta (7 -10 Hz) 120 ms

B

Zzzz

Ripple (200 Hz)

5 ms

C

Prefrontal cortex

Hippocampus

that these rhythms constitute a hierarchical system in the cerebral cortex. He places them along a logarithmic scale of frequency that reveals an astonishing property: the oscillatory frequencies are distributed according to regular intervals.[21] He suggests that increasing the frequency of the oscillations of the neuronal networks corresponds to the appearance of ever more complex brain functions, demanding ever greater processing speed. A very nice example of the role of oscillations in encoding space is the recently formulated hypothesis that the grids of the entorhinal cortex result from interference processes (such as visual interference) between oscillations intersecting at the level of the neuronal dendrites of the entorhinal cortex.[22] Here we find a fundamental aspect of simplexity: the use of time to encode space.

Another example of intervention of temporal encoding in cognitive function is the problem of long-term memory of events (or episodes). How, for example, does memory store each spatial trajectory? Many experiments show that it is indeed the entorhinal cortex and the hippocampus that encode places. But the hippocampus is the seat of rhythmic activity: "theta" rhythm, whose frequency is around 6 to 10 Hz. It has been shown that when a rat moves about, the successive points of its trajectory are encoded by place cells through a process known as phase precession, which is also associated with grid cells. However, grid cells are activated at precise moments of the cycle of

---

(Opposite) Figure 21. "Go to sleep. You'll remember it in the morning." (A) During exploration, a rat successively activates *place cells* in the hippocampus. The vertical bars represent the action potentials of these neurons. Place cells also give rise to slow oscillatory activity known as *theta rhythm*, whose frequency is around 6 to 10 cycles per second (Hz). Action potentials are produced at precise instances in the theta rhythm cycle, providing temporal encoding of the trajectory. (B) Following exploration, as the rat sleeps, brief (fraction-of-a-second) high-frequency oscillations (200 Hz), or ripples, cause the cells to discharge spontaneously in the same temporal order, as if the rat were "dreaming" its trip. The information is compressed as in modern telecommunications. (C) These ripples are echoed in the prefrontal cortex, whose neurons are synchronized with those of the hippocampus. This mechanism is believed to contribute to the transfer of memories constructed in the hippocampus toward the prefrontal cortex for long-term memory. [Adapted from A. Peyrache, M. Khamassi, K. Benchenane, S. I. Wiener, and F. P. Battaglia, "Replay of Rule-Learning Related to Neural Patterns in the Prefrontal Cortex during Sleep," *Nature Neuroscience*, 12 (2009): 919–926.]

an oscillating wave that modulates their membrane potential — theta rhythm. The phase of the moment when the place cell fires encodes the place and even the episode. This constitutes a relationship between time and place.

The findings revealed something even more surprising. According to Buzsáki and his colleagues, episodes appear to be transmitted to the prefrontal cortex for long-term memorization through the intermediary of very rapid ripples of activity (200 Hz) that are produced when the animal sleeps or is resting. If that is true, it means we have to rest often when we learn something to fix it in long-term memory! During these ripples, the activity of populations of neurons in the prefrontal cortex is synchronized and activated by ripples that are synchronized with those of the hippocampus[23] (see fig. 21). Although we do not understand all the details of these mechanisms, it is impossible not to be impressed by the elegance of the solutions. Sometimes, time is on the side of simplexity.

11. Perceiving, Experiencing, and Imagining Space

The preceding chapter presented examples of the variety and the importance of brain mechanisms dealing with or using space. We also touched upon the problem of the relation between space and time and how it may contribute to simplexity. Now we will explore some of the more challenging questions relating to the role of space. Specifically, we will consider the foundations of geometry, first because, as we have seen, the brain is structured according to geometric kinematic laws, and second because mathematicians specializing in geometry use the word "simplex."

The Concept of Simplex in Mathematics

When I first thought of coining the word "simplexity," I admit that I was unaware of the fact that mathematicians—in particular, geometers—use the concept of simplex.[1] In geometry, the word *simplex* designates shapes in *n* "dimensions" that generalize the triangle, which along with the circle is the simplest shape. Assembling or combining very elementary fragments or shapes known as simplexes gives rise to all possible geometric shapes. What is important here is the idea of deconstructing a form into similar elements whose only characteristic is their dimensionality. But careful! Dimension here is taken not

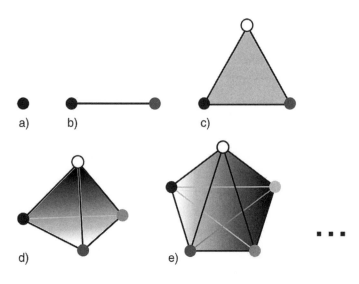

Figure 22. Simplex geometric shapes. Examples of geometric figures of various dimensions.

to mean amplitude but rather the number of required coordinates. However, a shape requires more information than a number; it requires a structure. Thus, a line segment will have one dimension, a plane triangle two, a tetrahedron in space three, and so forth[2] (see fig. 22).

The term *simplex* also belongs to affine geometry. Henri Poincaré was the first to invoke what are still known as "simplicial decompositions" in topology of any dimension, but he did not use the term *simplex*. He referred to "polyhedra." Recall that affine geometry has very general properties. Its distinctive feature is that it ignores incidence relations and preserves only properties relating to incidence and parallelism. The study of perspective illustrates this very well: From an affine point of view, two shapes are "identical" when one has a "cavalier perspective" (also known as a high viewpoint) of the other—the view from infinity, if you will. Under these conditions, a circle has the "same shape" as an ellipse when seen from a certain angle.[3]

The first person to systematically use the word *simplex* to designate these high-dimensional shapes vis-à-vis the generalization of triangles was probably the Dutchman Pieter Hendrik Schoute, around 1901.[4] The idea was to obtain the simplest shape possible while having access to *n* dimensions. The first person to define and use the expression *simplicial complex* was Schoute's compatriot Luitzen Egbertus Jan (frequently cited as L. E. J.) Brouwer in his demonstrations of invariance of dimensions.[5] It was probably he who introduced the term *simplex* in the discipline that today we call topology. At this stage, the word existed as did the methods of combinatory topology, but the explicit naming of a separate discipline had to wait for the later work of James Alexander.[6]

The Role of the Body in Sensory Experience

The above notions of simplex defined by mathematicians are very abstract. But are geometric concepts really totally abstract, or can they be linked to the organization of the brain and the body? In other words, can we ask whether the concept of simplex has a biological foundation? I suggest that the considerable simplification offered by abstract geometry in apprehending the world may reflect the principles of life itself. Here we are touching on a fundamental debate between those who believe that "abstraction came first" and those, to whom I belong, who think that "the act of a living organism came first." The first group includes philosophers, such as G.-G. Granger,[7] who have defended the primacy of abstraction as a basic component of mathematics and, particularly, of geometry. The formalist school of spatial thought is also illustrated by Peter Gärdenfors, who was interested in metaphor: "Perception produces mental representations of objects in a conceptual space. These representations can be used for metaphoric transfers in precisely the same way as the mental representations generated by linguistic inputs."[8] In reducing metaphor to a transfer of geometric structures, Gärdenfors shares with mathematician David Hilbert the idea that to think is to name. However, in so doing, he neglects (in my opinion) the essence of what geometric structure means for an acting body: A geometric structure is, in effect, always an "act."

In considering the major theories proposed by mathematicians and philosophers regarding space, one is also confronted by the opposition between the ideas of absolute space and relative space.[9] Newton separates them by interpolating the senses: "Absolute space, in its own nature, without relation to anything external, remains always similar and immovable," he writes. Inversely, "relative space is some movable dimension or measure of the absolute spaces; which our senses determine by its position to bodies; and which is commonly taken for immovable space."[10] The perception of forces is decisive for establishing the notion of relative space. Physicist James Clerk Maxwell was also one of the first to introduce the potential role of the perceiving body in developing ideas about space. Maxwell sketched "an epistemology of muscular effort," about which he wrote:

There are, as I have said, some minds which can go on contemplating with satisfaction pure quantities presented to the eye by symbols . . . in a form which none but mathematicians can conceive. . . . There are others who feel more enjoyment in following geometrical forms, which they draw on paper, or build up in the empty space before them. . . . Others, again, are not content unless they can project their whole physical energies into the scene which they conjure up. They learn at what a rate the planets rush through space, and they experience a delightful feeling of exhilaration. They calculate the forces with which the heavenly bodies pull at one another, and they feel their own muscles straining with the effort. . . . *To such men momentum, energy, mass are not mere abstract expressions of the results of scientific inquiry* [italics mine]. They are words of power, which stir their souls like the memories of childhood.[11]

What is crucial for our purposes is to understand how the notion of space simplifies mental processes of reasoning. Recall our two points of view. The first says that because space is an abstract notion, it is an important tool for logical reasoning and abstract memory. The second holds that space is a notion anchored in the acting body: Accordingly, its "incarnation" in the acting body—its "enactive" aspect—is what gives even the most abstract reasoning its obvious character. How is this point relevant to my theory of simplexity? It is relevant because action is an "immediate datum of consciousness," and the anchoring of notions of space in action bridges a gap between the abstract and the real. Do not misunderstand me: I do not deny the formidable power of abstraction or of abstract thinking. I simply maintain that if a fundamental connection exists between the abstract and the act, this "reflexive" power is generally accompanied by a substantial potential for efficiency. The debate thus opposes two extreme positions: One assumes the existence of abstract entities that are independent of the senses; the other attributes an essential role to experience and to the action of the perceiving, acting body. Note that several attempts have been made recently to link spatial schemas and abstract thought.[12]

In contrast with the purely abstract definitions of geometry, great mathematicians have proposed that the cognitive basis of geometry and the notion of space are anchored in perceived experience and the acting body. Poincaré and Albert Einstein argued strongly for taking sensate experience and action into account in the development of geometry. Poincaré links the concepts of space and action in space and attributes a fundamental role to grasping, capture, and locomotion, which he calls "truths of geometry." According to him, localizing an object in space simply means representing to oneself the movements that would be necessary to reach it.[13] Poincaré also emphasizes the muscular sensations that accompany movements. This idea, defended by twentieth-century French philosopher Jean Cavaillès and the "intuitionists,"[14] is also supported by Einstein: "Poincaré is right. The fatal error that a mental requirement, preceding all experience, is at the basis of Euclidean geometry is due to the fact that the empirical basis on which the axiomatic construction of Euclidean geometry rests has been forgotten. Geometry must be considered as a physical science whose utility must be judged by its relationship to sensate experience."[15] Recall, however, that Poincaré also maintained, paradoxically, that geometry was a convention and thus distinct from perceived space.

Einstein also supports Poincaré in discussing what constitutes our notion of space. According to Einstein, Poincaré rightly emphasized that we distinguish two sorts of changes in material objects: "changes of state" and "changes of position."[16] The latter can be corrected by arbitrary movements of the body. Because he proceeds to a definition of space based on two objects, as did Poincaré before him, and because he insists on the fundamental role of relative changes of position between two objects in the composition of space, Einstein can define space as the quintessence of quasi-rigid extensions of a body, which constitute the empirical basis of our conception of space. According to Einstein, in prescientific thought, the solid crust of the earth played the same role as the human body — that of a reference.

The importance of perceived experience to geometry was suggested recently by logician Giuseppe Longo consistent with the ideas of the influential mathematician Hermann Weyl: "The formal analysis of mathematical theories is only one of the elements required for functional analysis: it is in no way a sufficient condition. . . . It is the meaning, constructed on the basis of 'our experiential acts' that permits the mathematician to 'understand' a proof or to formulate conjectures, and to the logician to formulate the conditions sufficient to constitute a formal system, the end and not the beginning, of a conceptual exercise."[17]

These questions were also considered by geometer Bernard Teissier, for whom the concept of "line" is fundamental to thinking about space, the mathematical line being the result of identifying a "visual line" and a "vestibular line." Here, briefly, is his idea: The visual line is what is perceived by vision on a sheet of paper or in nature. It is geometric, but lacks movement in itself. It is a simple, static image. To perceive it as movement along a line requires a contribution from another sense, for example, the system of vestibular receptors of the inner ear that serve to assess head movements. Teissier's intuition is that the signals put out by these sensors "synchronize" the visual line with time. He calls this transformation "Poincaré-Berthoz isomorphism," an allusion to the work in our laboratory on vestibular perception of rotations and translations.[18]

Today, mathematician Daniel Bennequin, together with mathematician and motor control specialist Tamar Flash, hypothesizes, as I indicated earlier, a more general theory of combinations of affine and Euclidean geometry, applicable for both perception and for diverse motor processes. In this way, the brain would endow the static visual world with "virtual movement" by "projecting" onto images the activity of a sensory system — that is, the vestibular system, which specializes in movement — even in the absence of movement![19] In support of this idea, brain imaging has shown that just the sight of a drawing of a curved line can activate the areas of the cerebral cortex that encode visual movement.

But what do these mechanisms have to do with our theory of simplexity? A great deal! Cooperation of the visual system with the vestibular system gives the brain an a priori perception of potential movement contained in the shapes of the physical world. In the language of mathematicians, the brain constructs groups of transformations that make it particularly apt at processing movement and the dynamic aspects of interaction with the physical world and even other living creatures.

Observing development in infants, the renowned Swiss psychologist Jean Piaget was quick to criticize Poincaré, who, according to him, had constructed a "conventionalist" theory that assumed too much fine-grained mental activity in the early stages of developing intelligence about space. "Nevertheless," wrote Piaget, "like Poincaré, we shall not hesitate to speak of groups to designate the child's patterns to the extent that they can be reversed or corrected to bring them back to the initial point." Piaget mainly objected to Poincaré's neglect of the fact that these "groups . . . remain in the practical state for a long time before giving rise to mental constructions." [20] Philosopher Maurice Merleau-Ponty in turn would reproach Piaget for the "classic intellectualism" that led him to translate perception into the language of intellectual operations and to have missed the real richness of the world perceived by children. [21]

The Concept of "Affordance" and Gibson's Theories

Although it seems clear that the brain is a geometric machine, the problem of the geometry, or geometries, that it uses has not been resolved. For example, rather than assuming that the brain makes complex geometric "calculations," one might consider that it is capable of directly extracting the invariants useful for carrying out action from the flow of visual or sensory information. Gibson called these invariants "affordances." [22]

For example, when you stand in front of a door, you immediately gauge the likelihood of getting yourself through it without having to take out a meter stick to compare the width of your shoulders and that of the door. Similarly, when you come to a wall, you are able to judge very quickly whether you can jump over it. Moreover, as also mentioned above, to living creatures, every object has a functional meaning. A slipper can be a warm refuge for a man's foot, a toy for my poodle, Lolita, or a yummy meal for a moth. Jakob Johann von Uexküll saw accurately that any given object has several possible meanings depending on the needs and modes of functioning of this or that species. Applied here to our thesis, these ideas confirm that space is not a calculable or abstract entity but rather a support for arrangements of shapes and objects that are useful to each species and have particular *meaning for a given action*.

12. The Spatial Foundations of Rational Thought

Tom Thumb imagined a remarkably simplex solution to the most complex of problems: finding his way back in a forest that was not familiar to him. He stuffed his pockets with pebbles that he tossed along the route. Likewise, Ariadne offered a simplex solution to Theseus when he left to vanquish the Minotaur in the labyrinth: She gave him a ball of string to unroll as he went that showed him the way out. Over the course of evolution, the problem of finding, or refinding, one's way has given rise to numerous biological solutions. Desert ants use polarized sunlight. Rodents and other animals mark their routes with odors. Bees have astonishing mechanisms. Migrating birds and fish and the emperor penguins that follow long trips across the globe have still other means. Even before Ariadne or Tom Thumb, from the very beginning, humans have migrated across continents, gone hunting in unfamiliar terrain, and set sail on the high seas. But they always had to come home. Consequently, they were forced very early on to develop various cognitive strategies to memorize routes, find shortcuts, and orient themselves in space. They invented increasingly sophisticated equipment, from the astrolabe of sailors to the navigational electronics that come with our cellular telephones.

I have two hypotheses on this subject. The first is that mental tools developed throughout evolution to resolve multiple problems of wayfinding in space were also used for the highest cognitive functions: memory and reasoning, relations with others, and even creativity.[1] The second hypothesis is that the mental mechanisms for processing space make it possible to simplify many other problems faced by living organisms. As we have seen in the preceding chapters, space serves to simplify neurocomputation in many ways, whether through the geometry of the neurons and the most elementary proteins; the use of the helix, not only for the genome but also to distribute sensory cells on the cochlea; the distribution of signals in neuronal maps or in columns; or modification of the geometry of neural maps depending on whether the function is sensory or motor.

Here I would like to stress the use of space to simplify certain highly "cognitive" functions. Indeed, it seems to me that the neural bases of mental manipulation of spatial frames of reference (egocentric, allocentric, geocentric, heterocentric, near, and far) constitute the basis of rational thought, in particular, the human aptitude for geometry, reasoning, changing perspective, simultaneously processing different "points of view," and contingent logic.

Together with the brain, these neural bases also appear to enable emotions, interaction with others, intersubjectivity, and even empathy.[2] For example, historians Frances Yates[3] and Mary Carruthers[4] have shown how, beginning with the Greeks, space has been used in the "art of memory" to store or find objects, places, events, words, and concepts, but also to find new combinations, invent histories, and create associations. Architecture is another illustration of how space helps us to organize concepts and ideas. The great master of Roman Antiquity, John Scheid,[5] discovered a Roman manuscript that mentioned a way through the city that was intended to educate elite Romans and, particularly, foreigners about social customs. This document contains maxims and recommendations that are organized according to the map of a tour through the city of Rome. Each chapter is associated with a monument or a remarkable place. Perhaps this method of presentation helped students to memorize the contents of the document. In other words, using space is not only a simplex mechanism for the senses, as I said earlier; it is also a simplex tool for rational thought.

Is the Language of Space the Same in Different Cultures?

Anthropologists have studied the diversity of systems of frames of reference used by people all over the world to get their bearings in geographic space. For example, on some islands, the inhabitants rely on the mountains for orientation; on others, the rising sun; on still others, the orientation system may be based on a combination of elements borrowed from geography, cosmology, and sometimes from the male or female body. Such is the case of the Miraña, who have been studied by anthropologist Dimitri Karadimas.[6] The common house of the Miraña, called the *maloca*, is oriented according to geographical space and also a cosmology. Its architecture is based on the skeletal combination of a man and a woman engaged in lovemaking (see fig. 23). The woman is lying on the ground and the man is on top. The entrance corresponds to the female sex. The architecture of the house is a remarkable integration of all the various spaces that the brain must deal with and for which it has separate but cooperating neural networks. The Miraña house is embodied, incarnated, and placed in the cosmic and earthly world, as we are. Here, architecture, human behavior, thought, imagination, and cosmic reality are mixed. The uniform cover of the house and its use for human social interactions are extraordinary symbols of the "unity of man and the universe."

Similarly, as the late anthropologist Claude Lévi-Strauss showed, the village of Bororo is organized according to a very rigorous geometry. Women run the huts, which are arranged in a circular plan; men share a "common house" in the center of the circle. The circle itself is divided into two parts, and men born in one part cannot marry women from the same part, to safeguard the diversity of the family tree. The rules that underlie the social life of these people can be called simplex because they reduce the complexity of relationships by creating "structures" (whence the term "structuralism," to designate systematic research into these rules).

The caste system in India is another, related example. It is based on a very strict hierarchy that simplifies things for the human mind, and more specifically, it relies on a "vertical" hierarchy that uses space as a frame of reference. The system operates according to complex relations and integrates many functional specializations, ethnic groups, and languages within the general system. The actors in the system — people as well as animals — are both dependent and independent. The castes themselves are divided into subcastes and into lines that replicate the same global schema. In fact, this type of organization masks substantial flexibility, as evidenced, for example, by the inversion of the hierarchical order between Brahmins, who dominate in principle, and aristocrats. Inversions are also evident in the way the body is symbolized: The upper part is associated with the heavens and the lower part with the ground, but the inverse is also possible.[7] A similar hierarchy can be found in the West in the degree of perfection of creatures in the great chain of being of the ancient Greek philosopher Plotinus. All these cases call into play simplex processes that enable this or that group to resolve substantial difficulty posed by orientation in space, navigation, and the random displacements of the hunt, as well as commerce, the harvest, and war. This is not to say that the frames of reference are the simplest, but that they are the most "convenient," as Henri Poincaré would say.

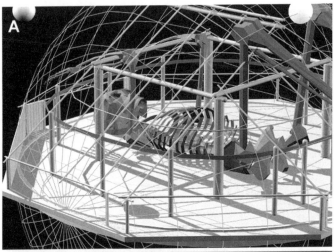

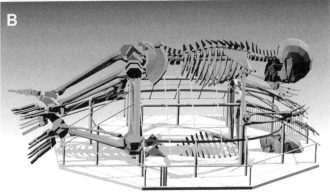

This diversity of frames of reference is reflected in the language used to verbally describe the location of objects in space. Three principle frames of reference are used around the world: a relative or egocentric frame of reference, which relies on point of view ("The ball is to my left"); an intrinsic frame of reference, referring to the object ("The ball is left of the tree"); and an absolute, or allocentric, frame of reference ("The ball is south of the tree"). Whereas in European languages, relative (egocentric) constructions dominate, in indigenous languages (Australia, Papua New Guinea, Mexico, Nepal) absolute (allocentric) ones do.

A recent study comparing frames of reference of inhabitants of a village in the Netherlands with those of a village in Namibia confirmed these generalizations.[8] The experiment required subjects to recognize the position of objects after a change in point of view. To understand the paradigm, you have to imagine that you are in front of a table on which five cups are sitting upside down. Someone hides an object under one of the cups while showing you which one; then you move to another place, and the first table is hidden behind a screen. Now you are in front of another table, symmetrical to the first table with respect to you. This other table has five new cups sitting on it, identical to the previous cups. Your task is to find the hidden object. Of course, three responses — that is, three strategies — are possible, depending on whether you encoded (and memorized) the position of the object on the first table in one of the three reference frames men-

---

(Opposite) Figure 23. Concepts and representations linked to space used by the Miraña people of the Amazon to construct their communal house (maloca). The house is built (A) according to a stellar cosmogony, directions keyed to geographic landmarks (rivers, hills), and (B) a sexed architecture inspired by the couple. The basic structure represents the backbone; chevrons, the ribs; and the entryway, the vagina. Here, architectural thought dissimulates a wealth of frames of reference linked to different worlds (body, geography, stellar space) in an apparently simple form. A Miraña house is simplex. To my mind, this also reflects the functioning of the brain, which creates nested worlds centered on the "acting body." [Adapted from D. Karadimas, *La Raison du corps. Idéologie du corps et représentations de l'environnement chez les Miraña d'Amazonie colombienne* (Paris: CNRS Editions, 2005).]

tioned above: relative (egocentric), intrinsic, or absolute (allocentric). For the Dutchman, the answer is egocentric; for the Namibian, it is allocentric.

Generally speaking, humans tend to use similar frames of reference for language and in carrying out spatial tasks. The experiment described above is interesting because it was also carried out with several species of nonhuman primates. It turned out that big apes had a preference for environmental encoding — "allo" rather than "ego." The researchers concluded that philosopher Immanuel Kant's hypothesis of priority of egocentric encoding and reasoning, also supported by Jean Piaget in the development of infants, was disputable: In humans, cognitive preferences in the choice of reference frames varies as a function of natural language preferences. This correlation is very strong from the age of 8 and persists to adulthood. According to linguist Stephen Levinson, adopting an egocentric strategy has a "cost," that is, it calls into play more complex processes. This is the reason infants acquire an "allo" strategy at 4 years of age and maintain it to around 8 years, whereas the "ego" strategy is acquired later, as we confirmed recently together with the child psychologist Olivier Houdé. It is also later, between 10 and 12 years, that a child acquires the capacity to "change point of view." I have proposed elsewhere that at this age, there is a "critical cognitive period for the capacity of changing point of view"[9] that is crucial for the acquisition of tolerance. If the child is put in a restricted social environment and is indoctrinated with only one point of view toward others, for instance, hatred of other religious beliefs or ethnic communities, he will develop a lifelong hatred and fanaticism.

In any case, all these experiences confirm the existence of cognitive strategies common to all cultures and to nonhuman primates.[10]

The Maze and the Garden

The maze and the garden are two universal examples of organization of space; they reflect the importance of shared space as a simplex tool in human thought. Although buildings of undetermined purpose but similar to a maze have been found in Egypt,[11] it is generally agreed that the form and realization of the maze was an invention of the Greeks, notably featured in the myth of the war between Theseus and the Minotaur. Everybody knows the story, but I will repeat it. To find the Minotaur and kill it, Theseus must negotiate a complex "labyrinth." Meeting this challenge, passing the test, achieving the impossible will make Theseus a hero. Once victorious, he will have to change his point of view and find his way back. Now, Theseus has only "egocentric" knowledge of the labyrinth, whose complexity makes it impossible to construct a mental map while navigating through it. Ariadne provides an elegant, simplex solution. In suggesting that Theseus use a string, Ariadne simplifies the problem and obviates the need to memorize the path.

It may be this uncertain and complex course that motivated monks to create mazes at the entry to churches and cathedrals. Take Chartres Cathedral. The problem appears to have been the following:[12] How could pilgrims entering the building be made aware that access to the divine is worth (and should come only at the end of) a lengthy approach? The labyrinth is the simplex solution to this problem. To get to the center of the cathedral, a pilgrim first must follow a sort of qualifying course that engages him in a mental process of decentering and manipulating perceived, experienced, and imagined space that renders him receptive to the mystical ambiance of the cathedral.

The qualifying course, at the entry to the church, reminds me of the "approaches" that led us (when I was still climbing the Mont Blanc range) to the hut after three or four hours of walking. The next day, at the stroke of two in the morning, we left to attempt the ascent.

At this hour, we realized that the approaches and the night in the hut had completely modified our connection to reality. We surrendered to the mountain, ready to be influenced by it, read its mysteries, compared ourselves with its force, and faced uncertainty and danger: The approaches had turned us into beings purified of our muddy layers, to paraphrase the sixteenth-century French poet Pierre de Ronsard. The maze at the entry to the church imposes a test of this sort, though naturally without the effort of the ascent. Still, it rattles one's linear, egocentric perception, and liberates, shifts, and manipulates the frames of reference.

The motif of the maze has endured through the ages. Many symbolic meanings have been attributed to it. The Tibetan *mandala* is one such example. For me, it represents the importance of space in structuring the major themes of life, and the subtle play between complexity and simplexity that is a major characteristic of life.

In contrast to its universality and the relative uniformity of a maze's geometry, gardens — which often contain mazes — illustrate the different ways of thinking among various cultures. The French garden is constructed in very straight lines, all geometric rigor, reflecting the Cartesian, formal, prescriptive thought that predominates in France. The English garden is all about variety and surprise. Here, nature is left to do the organizing, mixing impulse and error, shade and shadow, alleys and byways. It is the garden of a man of feeling. Indeed, it belies the implacable rigor of a provincial, stratified people who do not enjoy revolutions! Also worth mentioning among these notable landscapes is the Japanese garden. Neither simple nor complex, here and there, amid a carefully composed order masked by apparent whimsy, are rocks and streams, trees and flowers, like so many voices in a polyphonic song aimed at conveying an illusion of serenity. This garden is a model of simplexity and remarkably conducive to meditation. Like Japanese painting, where the empty spaces encourage the imagination to complete the strokes parsimoniously disposed on the canvas, the Japanese garden keeps the mind from getting lost in a welter of details and the simplicity of the trivial. It attains the highest degree of simplexity, which is not content with simplifying but leaves the observer the freedom to think creatively.

Ecumene Space

Space is not only what we move through in a maze or a garden, when we walk "around our room,"[13] or around our city. The brain does not only simulate our actual journeys or our mental tours of a city map. It also divides up space in many ways depending on the communities we are part of. Each of us lives in an apartment or a house, but also a city or a village, and then a state, region, country, continent. In Venice and in the grand medieval metropolises, occupations were often associated with specific sections of the city. These divisions, which aid social organization, favor simplexity. In this way, instead of being lost among several complex networks, the brain evolved easily among these different identities linked to religion, history, profession, or an individual's family.

A very fine simplex concept summarizes the richness of the human milieu and its varieties. It is the "ecumene," defined by French geographer Augustin Berque: "The ecumene is the condition of human milieux, with an emphasis on 'human,' but also ecological and physical aspects. It signifies the 'habitat' (*oikos*) of human beings. Taking this habitat into consideration . . . is to be opposed to the philosophy that would have localized the habitat of beings in language. It is also to oppose oneself to the too-human sciences which, in their way, assumed this part and, in so doing, separated culture from nature . . . even as they failed to negate the animality of our bodies! The effect of this contraction was once again to cut the human being in two, just as dualism had already cut him off from the things of existence."[14]

The complex problem posed by today's world can be expressed

in the following fashion. In olden times, an individual had relatively few identities, and the categories that comprised these identities were perennial, often linked to where he lived, his culture, his traditions, and his beliefs. A person might be a father or mother in their house, a worker at a factory, a lawyer in her office, a professor at a university, or a citizen in his country. People traveled little and emulated the world through books. One of the characteristics of the societies in which we previously lived was the rapidity with which we had to change spatial identity. We are frequently citizens in winter and tourists in summer during the peregrinations that make temporary immigrants of us. Our perception of the geography of the world is thus linked to our multiple life universes, which are interlocked, sometimes complementary, and now and then at odds. These universes change over the course of a life, even from one instant to another, for example, when we are on the phone.

Today, the brain must emulate worlds of incredible diversity—including the Web and such virtual worlds as Second Life: The spatial span of our lives varies almost inconceivably. One day, during a visit to a laboratory in Singapore, noticing the diversity of faces among the students in the big auditorium where I was speaking, I asked: "But where do all these students come from?" My colleague responded without understanding, "From the region!" I persisted: "But how big is the region?" My colleague smiled: "Five hours by airplane, never more." Here, space was encoded as hours in an airplane; in France, it is encoded in days or hours on a high-speed train: The scale is not the same. Each person finds his simplex scale; the problem of scale is yet another major challenge of life.

The risk of such a situation is that it could plunge a good part of humanity into despair, pitting those who know how to Google or to use a GPS, navigate between the real and the virtual, and switch identity and continent in just a few seconds thanks to a cellular telephone, against others who are crowded off the technology highway. For the latter to avoid becoming slaves of the former, we need to re-think (and urgently) categories of identity, which is not simple. It in-volves knocking down boundaries erected during the twentieth cen-tury between language and lived experience, reasoning and emotion, and the global and local. The stakes are that people may succumb to gurus offering them the bogus sanctuary of sects—as has already happened—paving the way for a return to obscurantism, creation-ism, and fanaticism.

Drawing in the Service of Literary Creation

Space is central to perception and action, memory and identity. But it also serves as a guide to creative thinking and narrative fiction by means of diagrams and drawing. The complexity of a great novel can be considerable—Tolstoy's *War and Peace,* Alessandro Manzoni's *The Betrothed,* and the monumental work of Marcel Proust, whose plot outlines are famous. Happily, this hard road to literary creation is made easier by a simplex, and handy, mental tool: drawing, in-valuable to the written description of complex social situations or for structuring a novel.[15] For example, in Stendhal's autobiographical *Life of Henry Brulard,*[16] a map comes to the aid of memory. Elsewhere in the book, Stendhal represents himself at the blackboard face-to-face with his schoolmaster. To convey the relationship of the two characters in the scene, Stendhal supplied a drawing that "in a single glance" enables the reader to "grasp" the situation better than a thou-sand words.

Like diagrams, drawings aid writers' thought processes.[17] No-where is this function more evident than at the banquet—the first big dinner party—given by the Rougon-Macquart family in Émile Zola's novel series of the same name. Charles Saccard assembles the actors in his diabolical plan to make a quick fortune through dubious speculation. Zola begins by drawing a table around which he places the names of the guests, just as one does in planning seating arrangements. At this stage, the sketch simply suggests spaces held for the guests, in the normal manner. But the table plan does not remain that way. The 28 guests swell to 38. The space becomes eroticized, and the table a fatal weapon of the seductresses. Another famous example, also from Zola, is a drawing representing a column of insurgents. The development of the space as supported by the sketch clearly shows that the topography of the novel transposes the geography of the region. Zola modifies the geography of the towns; in the drawing, they are aligned. Another drawing involves a family at a crossroads: Gervaise and her children. We see them take paths radiating out and leading to their individual destinies. Nana goes west, where she will seduce the men of the boulevards; Étienne heads north, where he will rejoin the Gouget ironworks; Gervaise will become a prostitute near la Villette, to the east. The plan both anticipates and precipitates the account of the family debacle. In the same novel, drawing is also used to construct a text in which passions dominate (see fig. 24).

Generally speaking, space and drawing are useful mnemonic techniques. They also help writers to manipulate point of view, that is, both change of perspective and change of opinion, when they need to describe the complex evolution of social relations. This is particularly clear in the case of Gustave Flaubert, who in *Hérodias* uses drawing to gradually construct his description of the situation of the fortress of Machaerus. "The citadel of Machaerus stood to the east of the Dead Sea on a conical peak of basalt. Four deep valleys surrounded it, one on each side, one in front, one behind. Houses crowded up against its base within the circle of a wall, which rose and fell with the uneven contours of the ground."[18] An initial drawing shows the view above the citadel: It provides a means of encoding the features of the site. Other, later drawings show the citadel in context, in its environment. Here, the drawing aids the text in expressing the power of the site not with the simplest words, but with the most appropriate ones.

It may also be that the use of space in literature is linked to the cognitive strategies that the writers prefer. Some might be more inclined to narrative discourse experienced in its temporal organization, preferring to describe a succession of events and places according to their appearance in time. This is a predominant property of the left brain, the brain concerned with language, details, memory of successive episodes in a life,[19] and of local scenes. Others might have a greater propensity for the big picture, topography, events—which is the province of the right brain, the brain that is concerned with the metric properties of space, global aspects, and emotions. In other words, writers choose the strategy that offers them the greatest simplicity, either because that is the way they like to tell a story or because they sense—not with knowledge of how the brain works but based on experience—that this mode of narrating is easier for the reader to understand.

Return to Reality

Having reached the end of this essay, since I dared to propose a new concept, it might be useful to look at some of its origins in the history of thought. Such a contextual account is naturally beyond the talents of a humble physiologist, like myself. Philosophers, historians of science, and historians of ideas would be able to do it; they might find my own efforts audacious, not to mention presumptuous. Accordingly, I will content myself with sketching a few broad outlines, specifying why it is important, and useful, to counter complexity with simplicity.

Philosopher Michel Foucault described the slow evolution of

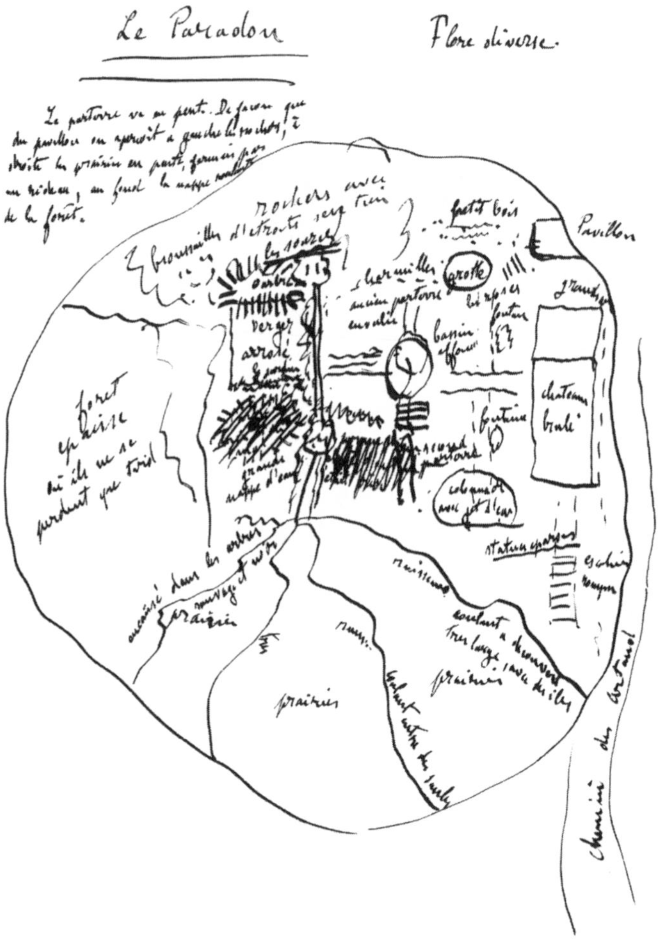
Le Paradou
Flore diverse
Le parterre va en pente. De façon que du pavillon on aperçoit à gauche les rochers, à droite les prairies en pente, fermées par un rideau, au fond la nappe de la forêt.
rochers avec
broussailles
la source
arbres
verger
arrosé
charmilles
ancien parterre envahi
petit bois
grotte
Pavillon
bassin
fontaine
château brulé
forêt épaisse où ils ne se perdent que tard
grande nappe d'eau
colonnade avec jet d'eau
statues éparses
escalier rompu
ruisseau
prairies
Chemin des Artaud

Western thought up to the end of the eighteenth century.[20] At that time, the world was seen as an abstract idea and gave rise to multiple interpretations. However, lacking any experimental science, much of the reflecting on the world was done using language as a principal tool. As a modern, twentieth-century thinker, Foucault emphasized the theoretical disarray that took hold at the end of the eighteenth century when classical thought and language "names, patterns, combines, and connects and disconnects things as it makes them visible in the transparency of words. . . . The profound vocation of Classical language has always been to create a . . . 'picture' . . . ; it has lost that secret consistency which . . . inspissated it into a word to be deciphered, and interwove it with all the things of the world; it has not yet acquired the multiple existence about which we question ourselves today."[21]

Henri Bergson, also a philosopher, and a magician with words, was also critical of language,[22] which he found inadequate to the task of describing the complexity, flow, and depth of the reality that makes up a human being: "We instinctively tend to solidify our im-

---

(*Opposite*) Figure 24. How drawing supports narrative thought: the Paradou. In *Les Rougon-Macquart*, Zola repeatedly made use of drawings to simplify the description of complex situations. For example, the Paradou is a mythical place of perdition and pleasure. It is synonymous with the love of Serge and Albine. Before writing the manuscript, Zola meticulously rethought the gradations of erotic desire. The spring is moved up a slope and made higher, as if to increase the force of the torrent. The tree will stand guard over the carnal union. Paradou thus becomes a rigorously constructed framework for a rite-of-passage narrative, told in stages. [Adapted from O. Lumbroso and H. Mitterand, *Les Manuscrits et les dessins de Zola: Notes préparatoires et dessins des Rougon-Macquart*, vol. 3, *L'Invention des lieux* (Paris: Textuel, 2002), by permission of the Bibliothèque nationale de France.]

pressions in order to express them in language. . . . Hence, we are now standing before our own shadow: we believe that we have analyzed our feeling, while we have really replaced it by a juxtaposition of lifeless states which can be translated into words."[23] Despite his dualism, Bergson deserves study today because of his intuitions and for having revived the notion of the act and of lived experience.[24]

Today, in our perpetual quest for "reality," we rediscover—after a certain fashion—its complexity. For reality is not easy to pin down. Jacques Bouveresse, a philosopher who specializes in Wittgenstein, declared: "I have made my peace with reality."[25] René Girard, historian and literary critic, suggests that many anthropologists have let structuralism blind them to reality, and in particular fundamentalist violence. Art historian Jacques Thuillier argues that art criticism should return to "phenomenology,"[26] akin to what French director Louis Jouvet would have called "the felt," in appreciating works of art, and that it should distance itself from formal theory. In giving renewed attention to the perceiving body and the role of emotion in reasoning, the cognitive sciences, too, are witnessing a similar movement.

But reality defies analysis. Although simplification has given rise to clearly observable uniformity among men, women, and even children—all consumers of Coca-Cola, all spectators of world championships in rugby and soccer, all fans of the same pop stars, all connected night and day by cell phone and the Web—we are also in the throes of a powerful desire to differentiate religions, traditions, languages, and ethnicities, fueled by all kinds of fanaticisms. The complexity of relations between society and nature seems beyond the ability of humans to deal with, whether from the vantage of the *Declaration of the Rights of Man and of the Citizen* or the United Nations. The violence that results is facilitated by the arms and drug trade. Reality—its brutality and its contradictions—makes us long for an ideal world, the world that existed before the eighteenth century, the one described by Foucault.

The Particular and the Universal: Ockham's Razor

The world is not simple. We can no longer control it. There is no God the creator who transcends an immutable reality, governed by predictable laws. The concept of hierarchy is being called into question at the level of the state, school, and businesses. Reason has run up against its limits. It is no accident that the 2002 Nobel Prize in Economic Sciences went to psychologist Daniel Kahneman (the award was shared with Vernon L. Smith), who showed that humans are not rational decision makers. From now on, probability and chance will frame the scientific study of life processes. Man and his world, or rather his worlds, are drowning in complexity and crave simplicity. The West is looking yearningly to Chinese philosophy, which is based on action.[27] Buddhism, a godless religion that allows each to find his way, attracts thinkers and Western scientists. The body is being celebrated in all its forms. The ideas of self-organization and autonomy in living things that were promulgated by neurobiologists Humberto Maturana and Francisco Varela,[28] centered on the concept of autopoiesis, have inspired the most advanced robotics.

Perhaps theories of complexity derive, in part, from the historical deception associated with any attempt to box life and nature. The theories aim at a new model of interaction that takes into account the infinite variety of forms and forces that constitute the relationships between the physical and living worlds. They also highlight the important discovery of the relationship between "chance and necessity."[29] Today, the concept of probability as applied to biology and the cognitive neurosciences is mature, thanks to the incredible success of Bayesian theories. Nothing is certain anymore. Everything is probable. Despite the lethal character of cancer, young people flirt with probability and continue to smoke cigarettes that come packaged with warnings that read, at least on French tobacco packets, "Smoking Kills."

The relationship between simple and complex is to some extent a relationship between the particular and the universal. This observation is not new. Beginning with the Greeks, and into the Middle Ages, it was the subject of the famous "quarrel of the universals,"[30] but it still occupies modern philosophers. What is at issue is whether the human mind can go from the particularity of the objective present, in its multiple and complex appearances and functions, to the intellectual formulation of the universality of categories of objects. The work of philosopher Alain de Libera contains an analysis of the way the human mind proceeds in simultaneously identifying the particularity of things in the world and conceiving their universal character: The horse that I see in front of me also belongs to the universal category "horse." De Libera analyzes the (to my mind, critical) contribution of thirteenth-century theologian John Duns Scotus, who renewed Aristotelian thought by placing at its center mental mechanisms that enable one both to grasp and to memorize the concrete object and its universal abstraction through the act of thinking, which makes possible the transition from one to the other: "In asking whether perception ['sensible things'] 'applied to man and not Callias the man,'" recalls de Libera, "Aristotle reverses Avicenna's proposition, affirming: I see the horse and not horseness. Through his theory regarding the intuitive knowledge of abstract acts of knowing, [John] Duns Scotus simultaneously explains how I see horseness, how I see the horse, and how I see that I see both horseness and the horse. The theory of universals and the theory of perception are articulated in a single theory of acts of knowing."[31]

And what of plurality? Must we take it into consideration or do we keep only the simplest features of aspects of reality? Everybody is familiar with the principle of Ockham's razor,[32] which we readily reduce to the idea that the best solution is often the simplest one. In fact, the idea probably stems from Aristotle ("plurality must not be used unnecessarily")[33] and the analogy to the razor that eliminates useless propositions from philosopher Étienne Bonnot de Condillac, the mathematician William Hamilton having later promoted it. In my opinion, the "principle of simplicity" attributed to William of Ockham is actually a "principle of simplexity." Ockham's idea is subtle: Abstract forms of thought — "concepts," "intentions," "similitudes" with the external world, "intellections" — are mental signs that there is no reason to distinguish the act itself from "intellection."[34] Thus, it is unnecessary to assume a special subjective quantity apart from the act. It suffices in itself, which leads to a principle of economy involving the primacy of the act and obviating the notion of representation as an independent entity. As novelist Pierre Alferi has written: "Representation in the imagination is not a medium between the mind and the thing. It is a real act, the image, that modifies the mind itself like a real quality and refers it back to the thing, in its absence, according to its sensory aspects. Of course, it is an act oriented to this sensory aspect, but even if we can call this orientation a representation, it is not an intermediary object, an intentional being, but the keeper of an act."[35]

Epilogue

We have come to the end.

Examining the concept of simplexity applied to life reveals an exceptional wealth of simplex mechanisms that have appeared over the course of evolution. We have proposed a number of principles as the basis of a theory of simplexity: the fundamental role of inhibition, specialization and modularity, anticipation, detours, cooperation, and redundancy. We suspect there are others. However, it is clear that we have only begun to glimpse the fundamental biological mechanisms that enabled simplexity to emerge in life. Those are for future science to decipher.

Our investigation leads to a compelling conclusion. We can understand nothing about life if we content ourselves with studying its building blocks. We must reintegrate these elements into their world, their *Umwelt*. We must restore the act to the heart of the molecule or the gene. Moreover, we need to create a kind of interactionism where the pressure of evolution to induce simplex solutions comes from the exchanges of living organisms with the physical world, whose properties they internalize to aid survival. The renewal of interest in epigenetics is testament to this necessity. Likewise, we cannot understand individuals without understanding their interactions with others, that is, intersubjectivity. The philosopher Paul Ricoeur formulates this idea very elegantly as "oneself as another."[1]

The rules and regulations of social life reflect the laws of perception and action. They also resemble those of simplex thought. The norms and rules of ethics and democracy appear to complicate things; in fact, they simplify, or at least make it possible to live in society. They attempt, often without success, to avoid mass murder.

The quest for simplexity in society confronts at least two major pitfalls. The first is illustrated by the example of economic theories, which for a hundred years have pretended to simplify economic life by regulating it according to mathematical models. At the end of the day, they have evolved toward a virtual universe totally divorced from the reality of people and of the perceiving subject. The second pitfall, in contrast, lies in the effort to reduce simplexity to a trivial phenomenology of lived experience by making a kind of compromise between the complexity of the world and simple solutions like those advertised on the Internet. Simplexity demands a lot more. It asks for innovation, invention, productive detours, selection, consideration of the past, and anticipation of the consequences of future action.

We have but opened a door. At the conclusion of this book, having attempted to combine speculations, hypotheses, and facts both proved and unproved, according to the deontology of science, allow me to take a step beyond the frontiers of science. Some will judge the exercise futile and, consequently, useless; others, more indulgent, will find it bold or at least (which would delight me) distracting. For the notion of simplexity, used by business, design, and engineering, can also serve in other areas of human activity. Architecture is only one example among many.

The Joy of Roofs and Staircases

The abuse of simplicity is evident in architects' attitudes toward buildings. The roof has a charm that they seem to have forgotten.[2] Since amnesia set in, builders have been covering our heads with flat roofs, which are simpler to design, covered with awful chimney stacks and the workings of elevators. They do not even have the savoir faire to make terraces that might encourage dreaming on summer nights and offer up the attractions of the urban landscape. Understand: I have nothing against flat roofs. The cities of the East are covered with them. But precisely because of its horizontality and its status as a sign of stability, a roof must be dignified. Whether it serves as a reminder of the horizon, which is difficult to glimpse from the depths of the city, or the interface of wall and sky, or is transformed by the setting sun into the border between day and night, a roof— even a flat one—can be exposed to the whims of the wind and light. It can evoke the terraces of Babylon and the streams of the Alhambra. It can offer shelter from a storm, like the upper deck of a ship-like building navigating the spaces of the galaxy. A roof can serve as an escape on days when the too-close quarters of the ship sets tempers aflare. It can be a place for celebrations whose joy radiates out over the city. A flat roof is a line on the horizon that emphasizes the elegant shape of a building. It reassures and, through its simplicity, sends a message of confidence, for it is perpendicular to gravity.

Throughout time, depending on the climate of their surrounds, their customs, or their beliefs, humans have varied the shape of roofs in puzzling ways. In Japan, roofs are superposed, in nested constructions. The inclined roof brushes the threshold of the house like an extended hand curving in a sign of welcome. A roof is short, like a skirt that opens to reveal lovely legs, or long, in countries where rain demands it and to allow snow to slide off naturally. A roof overhangs a wall to provide shade. Sometimes, the roofing of a house that shelters a farmer bulges, as though it held attics full of grain or other harvest. Sometimes, a roof is rounded or curved, depending on whim and custom. In Asia, for example, its comma points skyward. The roof is like a woman's hairdo. What woman would not attend to her hair? Styled hair is attractive; it emphasizes physical charms and rhythms; it is the link with the sky; it situates the face in the world and transforms its features.

The roof is like a hat. It may signal elegance or rusticity. We wear hats to protect ourselves from the cold and the sun; especially, we turn them into accessories. Some people must wear a hat in the presence of their God. Others take them off. To everyone, a hat is a symbol of their relationship with the cosmic powers. The hat is more than a parasol, umbrella, or windbreaker; it is reassuring. We manipulate hats in gestures that signal social relations: removing a hat in front of a lady or one's employer. Like a roof, a hat incorporates a simplex gesture.

Over the course of history, people also constructed stairs. Babylon had them. Stairs made it possible to mount horses in châteaus. Constructed in the form of a spiral, they wound up to the top of the tower of Sienna or led (with very long steps) to the summit of the Incan pyramids. They could be as discreet as concealed doors or as magnificent as the places they were leading to. The presentation of the virgin at the temple is almost always portrayed on a stairway. The front steps of the Élysée Palace in Paris, which one must climb to see the president, represent access to power, even today. Romantic or cavalier, authoritarian or refined, stairs are always a symbol, a gesture that accompanies and determines posture, rhythms, and relationships. They represent the transition between interior and exterior worlds. It is no accident that climbing the stairs constitutes the crowning moment at the Cannes Film Festival. However, this particular set of stairs strikes me as emblematic of our simplistic mediocrity. In the guise of architectural simplicity, the palace as a whole destroys the charm of the port and even the bay, one of the finest in the world. It undermines the ecumene of Provence. Equally awful is the ministry of finance at the Quai de Bercy in Paris, a bar thrown across the line of the Seine. On leaving Paris by way of the bridge of Austerlitz, with the elegant church of Notre Dame in the background, one would wish for a farewell gesture worth contemplating on the way to champagne country, to Alsace or Germany. Conversely, arrival in Paris should be met with a welcome that stimulates one's appetite for all the marvels the city contains. What one gets instead is a roofless building that looks like a barrier to a permanently closed tollbooth. It seems designed to humiliate people. I understand the desire to maximize surface area, but there are other ways to make a bridge that are not so disastrously simple.

One of the pleasures of walking in Paris is the variety of its roofs. You may have to wait until August for the streets to calm down enough for you to admire them. Look especially at the roofs at street corners. Often somewhat phallic in appearance, they are soaring domes punctuated with windows — the sort that invite dreaming or watching the sun rise. Hermit towers or sanctuaries for poets, they soften the angles of the buildings at the intersection of two streets, marking the places where several avenues meet. They rise like figures watching over the crossings where destinies converge. This imaginary dialogue is absent in today's buildings, where the angle is always simple.

The Magic of Street Corners

What about corners? The street corner is a place that should be dominated by simplexity. For lazy architects, street corners pose no problem. The angle of the building is dictated by the angle of the two streets, and thus the corner (in concrete) results from the principle of simplicity. But the corner is not that way at all. Look at Paris. Admire the variety of the corners. Some shelter bistros, others shops, others a fountain or maybe simply a bench. Why? Because in the olden days, architects understood that a corner is a special place. It is a place where one's perambulation changes direction. A corner cannot be too sharp. It must, by default, blend with the natural shape of the trajectory, which resembles a section of a spiral.

A corner may cut away, enabling one's gaze (which always anticipates change of direction) to avoid having to stop too suddenly in front of others. This kindness, this regard for the pedestrian, is an elementary courtesy. This sort of corner is also an ideal place for a meeting. Finally, it can be a poetic non-place. In each case, it bears witness to this subtle moment of decision — depending on whether one is in love or on the way to a business meeting — to leave together or to separate. These encounters must be helped along. Of course, a cutaway of the corner is a challenge for the architect in terms of story design, but it also gives rise to balconies and to bull's-eye windows that give the impression the building is watching the street. The window at the corner offers the freedom to look without having to follow the line of the avenue. It both sees and is seen. Its balcony offers a space for the imagination: You could put a garden there. The corner is the paradigm of architectural simplexity.

Music and Cerebral Lateralization

Let us leave architecture behind and move on to music. It is remarkable that all cultures have a limited number of notes that make up the combinatory repertoire that produces music. For me, this convention proves that the human mind tends toward simplexity. In proposing other forms of composing at the end of the nineteenth century, modern musicians may well have broken with classical music. But they still use a staff of ten lines. A very simple frame of reference underlies the complexity of musical combinations, and our brain clamors for simplexity.

Surprisingly, musicians have also adopted (at least in classical music) a two-part structure: that of the principal theme, played by an instrument or a group of instruments generally in the higher registers, and that of the complementary bass harmony, played in the low registers. This structure seems to me to correspond to the lateralized processing carried out by the brain. The theme is processed by the left brain, and the bass by the right brain. This division exactly replicates the distribution of functions between the left and right brain, the left being specialized in language, narrative, sequential events, whereas the right brain is concerned with context, particularly spatial context and emotion.

If the left brain prefers higher tones and the right brain lower ones, this corresponds to the fact that, in nature, the sound context is made of noises that, because of the spatial distribution, are filtered and, in general, are mostly situated in the low registers, whereas meaningful sounds (birds singing or dogs barking) are made in registers higher than natural sounds (falling water, rain, thunder, the wind in the leaves). This cerebral lateralization could explain why female voices are usually higher than male voices. Taking into account the correspondence that exists between perception and action, women (who use their left brains more) default to the high sounds that their brains prefer. If you will permit me another hypothesis, I would say that because high frequencies are easily filtered, they are more suited to short-distance communication, whereas male voices, which are lower, have a greater chance of being heard far away and correspond to the male aptitude for constructing allocentric spaces. One might likewise explain why, in the great orthodox liturgies, moments of great calm follow climactic moments. This alternation of calm and storm corresponds to two modes of cerebral functioning that complement each other and that, despite the apparently contrasting character of the two humors they incite, give the musical narrative a unity. Here, simplexity is achieved by the harmonious succession in time of two opposed and marvelously harmonious modes.

Consciousness: A Simplex Theory of Reality

The basis of our perception of the world and of ourselves is not only action, but the act — the act with its intentionality, its memory of the past, its projection onto the future, the specificity of what interests us in the world as a function of our *Umwelt*. The essence of the act is anticipation, and the brain is essentially a machine that anticipates by creating probabilities, that simulates reality before acting in a very brief delay that precedes action. It is absurd to understand the act as a unique event to which we can assign a before and an after. Every act is inscribed in the continuous flow of life. It is, as Husserl said, magnificently framed by protention and retention. Self-identity is also a flow between autobiographical memory, perception of the phenomenal self at each instant, and prediction of the future. The brain can be thought of as building a virtual world, a sort of dream lived by its double, which enables the continual simulation of the act and the choice of best solutions or inhibition of futile acts.

As proposed by Rodolfo Llinás, the brain is a closed machine that chooses the frames of reference, the effectors, and the entire repertoire of life to survive and adapt as a function of our *Umwelt*. This is why the brain can produce dreams and hallucinations that are actually modes of normal functioning that underlie our illusion of rationality. The human brain is a creator of worlds. Its reality is the processes it subtends. This mode of operation is essentially simplex because it frees our cerebral machine from the complex reality of the world by permitting it to navigate in this virtual simulation. We are aware when we dream, even though we are sleeping, even though we are disconnected from the external world. Consciousness is thus nothing other than this internal reality continually being updated to prepare, emulate, and simulate acts. It is a projector for consciousness that clarifies with its light.[3]

But this process is not unique. No doubt, a different state appears each time the brain anticipates an action. And because perception is always simulation of an action in the world, conscious perception is always anticipation of some event that will occur in the world, whether this event is produced by the perceiving subject or not. Awareness is not consciousness of what we are doing, given that we are aware after anticipating. Jean-Paul Sartre and Maurice Merleau-Ponty alike said that we are always aware "of something." I say that we are aware of things that we are anticipating. Awareness is one of the tools invented by evolution to allow us to anticipate while choosing. That makes it a simplex mechanism because it corresponds to the coherent construction of the reality of an instant, which may be useful to the brain depending on the goal and internal and external states of flow of the "perc-action." [4]

Love: The Ultimate Form of Simplexity?

A fundamental problem of evolution was to maintain stabilities and invariants: Perception of the invariance of an object, perception of our own body, of our identity, is evidence of that. However, living organisms also had to resolve the problem of stability in the relationships between individuals. To guarantee the safety and well-being of an infant, parental stability is fundamental and assumes attachment. Actually, an innate mechanism called imprinting ensures this attachment. We saw an example of it in the weeping camel. But this solution is not simple, and it carries risks. The problem is even greater in humans, who are continually driven by sexual desire. Some people get around the difficulty through strict social rules or by spatially limiting interactions. Today we know that borderline personality disorder, which causes seriously disturbed social and other behaviors, is partially a result of parental instability or early relationships with parents.[5] Is love the most elegant invention for ensuring the stability of a relationship that endures beyond the sexual act? In uniting lovers, might it create a link a thousand times stronger than any social norm?

A Little Potpourri in the Form of an Afterword

It might one day be amusing to write a dictionary of the terms relevant for simplexity in the different domains where it applies. In terms of language, it is clear that grammar, whether one is interested in its diversity or its generative and universal character, is a simplex process. Metaphor, too, is a wonderful way of shortcutting language. Mottos, logos, and acronyms are all manifestations of the desire to simplify without totally abandoning complexity. When I say "spark a conversation," I sum up in just a few words many actions and perceptions that would be complicated to explain. This précis, which would seem to diminish information, actually imbues the thing or process with vitality and richness. Similarly, the concept of *myth,* a detour through the imaginary, implies realities and intricate relationships that can be digested despite their apparent complexity.[6] Fables are another marvelous simplexity. The "Find somewhere else to sing!" of the ant is far more telling than the endless chatter of the grasshopper.

Simplexity appears in symbolic figures like the seven dwarves. Dopey is not as dopey as he seems. He is like Lou Ravi — the Christmas crèche figure popular in Provence — not a village idiot but a witness and wise presence, a wonderer. Modern society appears to have no place for such creatures. Yet as Michael Edwards,[7] at the Collège de France, has written, wonder is one of the highest forms of simplexity. To be in a state of wonder is not to be stupefied with rapture. Rather, it is to enter into a state of joyous readiness, open to any possibility. Great sages, and many researchers, are all wonderers.

Sometimes a simplex figure is an invention. This is the case of the Irish leprechaun, a fairylike creature that is a complex mix of good and evil. Similarly, in Burma, the faithful, whose fervor can be immense, do not revere Buddha only but also a series of little figures, *Nats*, each one of whom represents a component of the forces of nature in action. My friend Eamon Connolly gave me a fine example of a simplex explanation invented by fishermen in the west of Ireland. When I went fishing with him in Connemara, a land of fairies and sirens, he told me once we were several kilometers out to sea that there were mackerel under our boat. I asked him how he knew, and he said his *gnach* had told him. Indeed, we had a phenomenal haul. For the Irish, the *gnach* represents a complex set of clues — the presence of seagulls, the condition of the sea, the season, the hour of the day, the tide, or the location — that constitute the collective memory and wisdom of the fishers in a particular place. To some extent, it is intuition. For humans, the invention of divinities represents an exceptional simplex strategy for relating to nature. But now I will stop at the risk of being accused of seeing simplexity everywhere.

Simplexity Is . . .

Simplexity is a way of living with the world. It favors elegance over dullness, intelligence over formal logic, subtlety over rigidness, diplomacy over authority. Simplexity is Florentine; it anticipates rather than reacts, suggests laws and interpretive grids, is forbearing. It is adaptive rather than normative or prescriptive, probabilistic rather than deterministic. It takes into account the perceiving body as well as consciousness, and it considers context. Simplexity is intentional. It economizes energy, but sometimes it consumes it. It remembers time past, allows changes in perspective, is creative, and promotes tolerance, which is opinion with a lid on. That is what simplexity means to me; what it means to you is for you to decide.

Notes

Preface

Epigraph. A. Einstein, "Ernst Mach," *Physikalische Zeitschrift*, 17 (1916): 101–102.

1. M. Gell-Mann, *The Quark and the Jaguar* (New York: Henry Holt, 1994).

2. N. Johnson, *Two Is Company, Three Is Complexity* (Oxford: One World, 2007).

3. See G. Gigerenzer et al., *Simple Heuristics That Make Us Smart* (New York: Oxford University Press, 1999).

4. A. Gupta, J. Lee, and R. J. Koshel, "Design of Efficient Light Pipes for Illumination by an Analytical Approach," *Applied Optics*, 40 (22) (2001): 3640–3648.

5. W. Klonowski, "Simplifying Principles for Chemical and Enzyme Reaction Kinetics," *Biophysical Chemistry*, 18 (2) (1983): 73–87.

6. In German, *simplex* describes the root of a word minus any affix; it thus contrasts with the word *complex*, which signifies "composed."

7. Gell-Mann, *The Quark and the Jaguar*, 45. Note that the "plic" for the folding and the "plex" for the braiding both come from the same Indo-European root, "plek."

8. N. Bellomo, *Modeling Complex Living Systems: Kinetic Theory and Stochastic Game Approach*, Birkhäuser XII (Boston: Birkhäuser, 2008). Bellomo elaborates the ideas developed by L. H. Hartwell, J. H. Hopfield, S. Leibler, and A. W. Murray, "From Molecular to Modular Cell Biology," *Nature*, 402 (6761 suppl.) (1999): 47–52.

9. Theologian Nikolaus Cusanus (ca. 1401–1464) also thought about the problem. Inspired by the commentary of the Neoplatonist philosopher Proclus on Plato's *Parmenides*, the *Institutio geometria* of sixth-century Roman philosopher Boethius, and English scholar Thomas Bradwardine's *Geometria speculative*, Cusanus wrote his *De docta ignorantia* (On erudite ignorance) after 1444 (critical edition by R. Klibansky and E. Hoffmann, 1932). In it, he proposed to resolve the problem of simplicity/complexity by what he called the *coincidentia oppositorum*, the "coincidence of contrary forces or things." His method for so doing was the two quadratic pyramids based on Pythagoras's *Table of Opposites*.

10. In general, Leibniz begins with a Cartesian idea. In the expression *Nihil est in intellectu quod non prius fuerit in sensu* (Nothing is in the mind that was not first in the senses), wrongly attributed by thirteenth-century theologian John Duns Scotus to Aristotle, Leibniz adds the phrase *nisi ipse intellectus* (if not his own mind). Thus, his position was contrary to that of his contemporary, philosopher John Locke, suggesting that the mind or the brain contains the principles on which diverse ideas are based. Elaborating this proposition, Leibniz wrote *Theoria motus concreti* (in *Opuscules et fragments inédits de Leibniz*, ed. L. Coutourat [Paris: Alcan, 1903]), where he develops the idea of the ether as a universal cause of movement that can mechanically explain the diversity of phonemona in the visible and sensory world. This notion, again essentially Cartesian, contrasts with the atomistic physics of British philosopher-scientist Francis Bacon and his French intellectual peer Pierre Gassendi. In both cases, Leibniz keeps the idea of simplicity in complexity — an idea that, for him as for Descartes, depends on mathematics and the natural laws not visible to the observer. Leibniz uses a parallel principle to explain the invention of the alphabet which, according to him, enabled the truths of reason in any area of knowledge to be known, at least approximately, through calculation, as in arithmetic or algebra (Brian Stock, personal communication).

Part I: Remember to Dare

Epigraph. C. von Clausewitz, *On War*, trans. M. Howard and P. Paret, vol. 1, chap. 6 (Princeton, NJ: Princeton University Press, 1976).

Chapter 1: Making the Complex Simplex

Epigraph. U. Alon, "Simplicity in Biology," *Nature*, 446 (7135) (2007): 497; see also R. Milo, S. Shen-Orr, S. Itzkovitz, N. Kashtan, D. Chklovskii, and U. Alon, "Network Motifs: Simple Building Blocks of Complex Networks," *Science*, 298 (5594) (2002): 824–827; U. Alon, "Network Motifs: Theory and Experimental Approaches," *Nature Review of Genetics*, 8 (6) (2007): 450–461; R. Milo, S. Itzkovitz, N. Kashtan, R. Levitt, S. Shen-Orr, I. Ayzenshtat, M. Sheffer, and U. Alon, "Superfamilies of Evolved and Designed Networks," *Science*, 303 (5663) (2004): 1538–1542. See also W. Klonowski, "Simplifying Principles for Chemical and Enzyme Reaction Kinetics," *Biophysical Chemistry*, 18 (2) (1983): 73–87.

1. J. von Uexküll, *Umwelt und Innenwelt der Tiere* (Berlin: Springer, 1909); A. Berthoz and Y. Christen (eds.), *Neurobiology of "Umwelt"* (Berlin: Springer, 2009).

2. U. Alon, "Simplicity in Biology," 497.

3. U. Alon, *An Introduction to Systems Biology: Design Principles of Biological Circuits* (Boca Raton, FL: Chapman and Hall/CRC, 2006); M. Ptashne and A. Gann, *Genes and Signals*, (Cold Spring Harbor, NY: Cold Spring Harbor Laboratory Press, 2002); H. A. Simon, *The Architecture of Complexity in the Sciences of the Artificial* (Cambridge, MA: MIT Press, 1996); M. A. Savageau, *Biochemical Systems Analysis: A Study of Function and Design in Molecular Biology* (Reading, MA: Addison-Wesley, 1976).

4. D. J. Watts and S. H. Strogatz, "Collective Dynamics of 'Small-World' Networks," *Nature*, 393 (6684) (1998): 440–442; N. Nathias and V. Gopal, "Small Worlds: How and Why," *Physics Review E*, 63 (2) (2001): 021117; S. Wuchty, "Small Worlds in RNA Structures," *Nucleic Acids Research*, 31 (3) (2003): 1108–1117; P. Kourilsky, "Quality Control of Immune Self Non-Self Discrimination," in *The Biology of Complex Organisms: Cre-*

ation and Protection of Integrity, ed. K. Eichmann (Basel: Birkhäuser, 2003), 53–59.

5. The mathematician René Thom suggested an interesting approach: Focus on singularities, bifurcations, and other critical elements, and simplify the rest as much as possible.

6. J. Droulez and A. Berthoz, "A Neural Network Model of Sensoritopic Maps with Predictive Short-Term Memory Properties," *Proceedings of the National Academy of Sciences of the United States of America*, 88 (21) (1991): 9653–9657.

7. A. Berthoz, A. Grantyn, and J. Droulez, "Some Collicular Efferent Neurons Code Saccadic Eye Velocity," *Neuroscience Letters*, 72 (3) (1986): 289–294.

8. R. R. Llinás, *I of the Vortex: From Neurons to Self* (Cambridge, MA: MIT Press, 2001).

Chapter 2: Sketching a Theory of Simplexity

Epigraph. M. Merleau-Ponty, *The Structure of Behavior*, trans. A. Fisher (Boston: Beacon Press, 1963), 146–148.

1. A. Diamond, "The Interplay of Biology and the Environment Broadly Defined," *Developmental Psychology*, 45 (2009): 1–8; A. Diamond and D. Amson, "Contributions of Neuroscience to Our Understanding of Cognitive Development," *Current Directions in Psychological Science*, 17 (2008): 136–141; A. Diamond, N. Kirkham, and D. Amso, "Conditions under Which Young Children Can Hold Two Rules in Mind and Inhibit a Pre-Potent Response," *Developmental Psychology*, 38 (2002): 352–362; D. A. Rennie, R. Bull, and A. Diamond, "Executive Functioning in Preschoolers: Reducing the Inhibitory Demands of the Dimensional Change Card Sort Task," *Development Neuropsychology*, 26 (2004): 423–443.

2. G. Leroux, J. Spiess, L. Zago, S. Rossi, A. Lubin, M. R. Turbelin, B. Mazoyer, N. Tzourio-Mazoyer, O. Houdé, and M. Joliot, "Adult Brains Don't Fully Overcome Biases That Lead to Incorrect Performance during Cognitive Development: An fMRI Study in Young Adults Com-

pleting a Piaget-Like Task," *Developmental Science,* 12 (2009): 326–338; G. Leroux, M. Joliot, S. Dubal, B. Mazoyer, N. Tzourio-Mazoyer, and O. Houde, "Cognitive Inhibition of Number/Length Interference in a Piaget-Like Task in Young Adults: Evidence from ERPs and fMRI," *Human Brain Mapping,* 27 (2006): 498–509; O. Houde and N. Tzourio Mazoyer, "Neural Foundations of Logical and Mathematical Cognition," *Nature Reviews Neuroscience,* 4 (2003): 507–514.

3. A. Berthoz, *Emotion and Reason: The Cognitive Science of Decision Making* (Oxford: Oxford University Press, 2006).

4. See the exhaustive review on anticipation and prediction in the visual system: K. Kveraga, A. S. Ghuman, and M. Bar, "Top-Down Predictions in the Cognitive Brain," *Brain and Cognition,* 65 (2007): 145–168.

5. J.-P. Changeux, *The Physiology of Truth: Neuroscience and Human Knowledge* (Cambridge, MA: Harvard University Press, 2004).

6. S. Hanneton, A. Berthoz, J. Droulez, and J.-J. Slotine, "Does the Brain Use Sliding Variables for the Control of Movements?" *Biological Cybernetics,* 77 (1997): 381–393. Bruno Siciliano and Jean-Jacques Slotine have also proposed solutions of task-related control for manipulator robots that simplify the control of redundant systems: B. Siciliano and J.-J. Slotine, "A General Framework for Managing Multiple Tasks in Highly Redundant Robotics Systems," *IEEE Proceedings ICRA,* 2 (1991): 1211–1216.

7. J. Droulez and C. Darlot, "The Geometric and Dynamic Implications of the Coherence Constraints in the Three-Dimensional Sensorimotor Interactions," *Attention and Performance XIII,* ed. M. Jeannerod (Hillsdale, NJ: Lawrence Erlbaum), 495–526.

8. I do not allude directly to the theory of finalism as developed, for example, by philosopher Raymond Ruyer, although I cannot deny that it crossed my mind in writing this.

Chapter 3: Gaze and Empathy

Epigraph. F. Cheng, *Cinq méditations sur la beauté* (Paris: Albin Michel, 2006), 108.

1. N. Chomsky, *Language and Thought* (Wakefield, RI: Moyer Bell, 1993).

2. C. Hagège, *Le Souffle de la langue. Voies et destins des parlers d'Europe* (Paris: Odile Jacob, 2008).

3. Llinás, *I of the Vortex*, 15.

4. In this regard see my chapter on gaze in A. Berthoz, C. Andres, C. Barthélémy, J. Massion, and B. Rogé, *L'Autisme. De la recherche à la pratique* (Paris: Odile Jacob, 2005), 250–294.

5. I addressed this question in *The Brain's Sense of Movement* (Cambridge, MA: Harvard University Press, 2000), 352.

6. J. B. Listing, *Vorstudien zur Topologie* (Göttingen, Germany: Vandenhoeck and Ruprecht, 1848).

7. J. B. Listing, *Der Census räumlicher Complexe oder Verallgemeinerung des Euler'schen Satzes von den Polyedern* (Göttingen, Germany: Dieterichschen Buchhandlung, 1862); E. Breitenberger and J. B. Listing, in *History of Topology*, ed. I. M. James (Oxford: Oxford University Press, 1999), 909–924.

8. R. A. Clark, J.-M. Miller, and J.-L. Demer, "Three-Dimensional Location of Human Rectus Pulleys by Path Inflections in Secondary Gaze Positions," *Investigative Ophthalmology and Visual Science,* 41 (2000): 3787–3797; R. A. Clark, J.-M. Miller, and J.-L. Demer, "Location and Stability of Rectus Muscle Pulleys. Muscle Paths as a Function of Gaze," *Investigative Ophthalmology and Visual Science,* 38 (1997): 227–240.

9. M.-H. Grosbras, U. Leonards, E. Lobel, J.-B. Poline, D. Le Bihan, and A. Berthoz, "Human Cortical Networks for New and Familiar Sequences of Saccades," *Cerebral Cortex,* 11 (2001): 936–945.

10. Neuroscientist György Buzsáki, of Rutgers University, proposed that the transmission of memories that are composed in the hippocampus is made through neuronal ripples that transmit to the prefrontal cortex the information necessary for memorization during sleeping or rest. G. Buzsáki, *Rhythms of the Brain* (Oxford: Oxford University Press, 2006); M. B. Zugaro, L. Monconduit, and G. Buzsáki, "Spike Phase Precession Persists after Transient Intrahippocampal Perturbation," *Nature Neuroscience,* 8 (2006): 67–71.

11. L. P. Acredolo, A. Adams, and S. W. Goodwyn, "The Role of Self-Produced Movement and Visual Tracking in Infant Spatial Orientation," *Journal of Experimental Child Psychology*, 38 (1984): 312–327.

12. A. Berthoz, "The Role of Gaze in Compensation of Vestibular Dysfunction: The Gaze Substitution Hypothesis," *Progress in Brain Research*, 76 (1988): 411–420; A. Berthoz and G. Melvill-Jones, *Adaptive Mechanisms in Gaze Control* (Amsterdam: Elsevier, 1985); A. Berthoz, "Coopération et substitution entre le système saccadique et les réflexes d'origine vestibulaire: faut-il réviser la notion de réflexe?" *Revue neurologique*, 145 (1989): 513–526.

13. P. Rochat, *The Self in Infancy: Theory and Research* (Amsterdam: North Holland, 1995).

14. T. Farroni, E. M. Mansfield, C. Lai, and M. H. Johnson, "Infants Perceiving and Acting on the Eyes: Tests of an Evolutionary Hypothesis," *Journal of Experimental Child Psychology*, 85 (2003): 199–212.

15. A. Berthoz, "Physiologie du changement de point de vue," in A. Berthoz and G. Jorland, *L'Empathie* (Paris: Odile Jacob, 2004), 251–275.

16. Some psychiatric patients have the impression that gaze is in the environment, a phenomenon that is described well in A. Dubois-Poulsen, C. G. Lairy, and A. Remond, eds., *La Fonction du regard* (Paris: Inserm, 1971), 363–368.

17. B. Wicker, D. I. Perrett, S. Baron-Cohen, and J. Decéty, "Being the Target of Another's Emotion: A PET Study," *Neuropsychologia*, 41 (2003): 139–146; C. Keysers, B. Wicker, V. Gazzola, J.-L. Anton, L. Fogassi, and V. Gallese, "A Touching Sight: SII/PV Activation during the Observation and Experience of Touch," *Neuron*, 42 (2004): 335–346; T. Allison, A. Puce, and G. McCarthy, "Social Perception from Visual Cues: Role of the STS Region," *Trends in Cognitive Sciences*, 4 (2000): 267–278.

18. See the chapter on gaze in Berthoz et al., *L'Autisme*.

19. C. M. Schumann and D. G. Amaral, "The Human Amygdala in Autism," in *The Human Amygdala*, ed. P. J. Whalen and E. A. Phelps (New York: Guilford, 2009), 362–381.

Chapter 4: Attention

Epigraph. H. E. Pashler, *The Psychology of Attention* (Cambridge, MA: MIT Press, 1998), 399.

1. G. Horn, "Pathways of the Past: The Imprint of Memory," *Nature Reviews Neuroscience*, 5 (2004): 108–120. This review shows that the mechanisms of imprinting also depend on emotional and motivational factors.

2. C. M. Moore and J. M. Wolfe, "Getting Beyond the Serial/Parallel Debate in Visual Search: A Hybrid Approach," in *The Limits of Attention: Temporal Constraints on Human Information Processing*, ed. K. Shapiro (Oxford: Oxford University Press, 2001), 178–198.

3. T. Shallice, *From Neuropsychology to Mental Structure* (Cambridge, UK: Cambridge University Press, 1988).

4. Husserl refers to sensory "hyle." Noetic syntheses process these data, and their correlates are objects and objectal properties. Noemas are intentional targets, but their filling is sensory.

5. W. James, *A Textbook of Psychology: Briefer Course* (New York: Holt, 1892), 84.

6. James, *Textbook of Psychology*, 38.

7. H. von Helmholtz, *Handbuch der Physiologischen Optik*, 2nd ed. (Hamburg, Germany: Voss, 1896).

8. James, *Textbook of Psychology*, 304.

9. Pashler, *Psychology of Attention*.

10. F. W. Mast, A. Berthoz, and S. J. Kosslyn, "Mental Imagery of Visual Motion Modifies the Perception of Roll-Vection Stimulation," *Perception*, 38 (2001): 945–957.

11. A. L. Paradis, J. Droulez, V. Cornilleau-Pérès, and J.-B. Poline, "Processing 3D Form and 3D Motion: Respective Contributions of Attention-Based and Stimulus-Driven Activity," *NeuroImage*, 43 (2008): 736–747; H. Peuskens, K. G. Claeys, J. T. Todd, J.-F. Norman, P. Van Hecke, and G. A. Orban, "Attention to 3D Shape, 3D Motion, and Texture in 3D Structure-from-Motion Displays," *Journal of Cognitive Neuroscience*, 16 (2004): 665–682.

12. J. Driver and C. G. Baylis, "Attention and Visual Object Segmentation,"

in *The Attentive Brain*, ed. R. Parasuraman (Cambridge, MA: MIT Press, 1998).

13. R. H. Khonsari, E. Lobel, D. Milea, S. Lehéricy, C. Pierrot-Deseilligny, and A. Berthoz, "Lateralized Parietal Activity during Decision and Preparation of Saccades," *NeuroReport*, 18 (2007): 1797–1800; D. Milea, E. Lobel, S. Lehéricy, P. Leboucher, J.-B. Pochon, C. Pierrot-Deseilligny, and A. Berthoz, "Prefrontal Cortex Is Involved in Internal Decision of Forthcoming Saccades," *NeuroReport*, 18 (2007): 1221–1224; D. Milea, S. Lehéricy, S. Rivaud-Péchoux, H. Duffau, E. Lobel, L. Capelle, C. Marsault, A. Berthoz, and C. Pierrot-Deseilligny, "Antisaccade Deficit after Anterior Cingulate Cortex Resection," *NeuroReport*, 14 (2003): 283–287; D. Milea, E. Lobel, S. Lehéricy, H. Duffau, S. Rivaud-Péchoux, A. Berthoz, and C. Pierrot-Deseilligny, "Intraoperative Frontal Eye Field Stimulation Elicits Ocular Deviation and Saccade Suppression," *NeuroReport*, 13 (2002): 1359.

14. M. N. Shadlen and W. T. Newsome, "Neural Basis of a Perceptual Decision in the Parietal Cortex (Area UP) of the Rhesus Monkey, *Journal of Neurophysiology*, 86 (2001): 1916–1936.

15. D. Sauto and D. Kerzel, "Evidence for an Attentional Component in Saccadic Inhibition of Return," *Experimental Brain Research*, 195 (2009): 531–540; K. L. Possin, J. V. Filoteo, D. D. Song, and D. P. Salmon, "Space-Based but Not Object-Based Inhibition of Return Is Impaired in Parkinson's Disease," *Neuropsychologia*, 47 (2009): 1694–1700; D. Qin and F. Zhengzhi, "Deficient Inhibition of Return for Emotional Faces in Depression," *Progress in Neuropsychopharmacological and Biological Psychiatry*, 23 (2009): 921–932.

16. For a detailed and instructive account, see S. Hochstein and M. Ahissar, "View from the Top: Hierarchies and Reverse Hierarchies in the Visual System," *Neuron*, 5 (2002): 791–804.

17. J. Moran and R. Desimone, "Selective Attention Gates Visual Processing in the Extrastriate Cortex," *Science*, 229 (1984): 782–784.

18. J. Braun, "Divided Attention: Narrowing the Gap between Brain and Behavior," in Parasuraman, *Attentive Brain*.

19. Attentional blinking is currently the focus of much literature. To cite only a few of the most recent articles on the subject: M. Adamo and S. Ferber, "A Picture Says More Than a Thousand Words: Behavioural and ERP Evidence for Attentional Enhancements Due to Action Affordances," *Neuropsychologia*, 47 (2009): 1600–1608; P. Craston, B. Wyble, S. Chennu, and H. Bowman, "The Attentional Blink Reveals Serial Working Memory Encoding: Evidence from Virtual and Human Event-Related Potentials," *Journal of Cognitive Neuroscience*, 21 (2009): 550–566; G. Hein, A. Alink, A. Kleinschmidt, and N. G. Müller, "The Attentional Blink Modulates Activity in the Early Visual Cortex," *Journal of Cognitive Neuroscience*, 21 (2009): 197–206; A. N. Landau and S. Bentin, "Attentional and Perceptual Factors Affecting the Attentional Blink for Faces and Objects," *Journal of Experimental Psychology: Human Perception and Performance*, 34 (2008): 818–830.

20. M. M. Chun and M. C. Potter, "A Two-Stage Model for Multiple Target Detection in Rapid Serial Visual Presentation," *Journal of Experimental Psychology: Human Perception and Performance*, 21 (1995): 109–127. Note in *Journal of Experimental Psychology: Learning, Memory, and Cognition*, 24 (1998): 979–992.

21. R. Parasuraman, J. S. Warm, and E. Judi, "Brain Systems of Vigilance," in Parasuraman, *Attentive Brain*.

22. J. Panksepp, *Affective Neuroscience: The Foundations of Human and Animal Emotions* (New York: Oxford University Press, 1998).

23. D. R. Gitelman, A. C. Nobre, T. B. Parrish, K. S. LaBar, Y. H. Kim, J. R. Meyer, and M. M. Mesulam, "A Large-Scale Distributed Network for Covert Spatial Attention: Further Anatomical Delineation Based on Stringent Behavioral and Cognitive Controls," *Brain*, 122 (1999): 1093–1106.

24. G. Bush, P. Luu, and M. L. Posner, "Cognitive and Emotional Influences in Anterior Cingulate Cortex," *Trends in Cognitive Science*, 4 (2000): 215–222.

25. One of their limits is that they do not take into account the fact that the receptor fields of the neurons in the highest regions of the dorsal and

ventral pathways are extremely large, which contradicts the idea of a focalization of attention on a little region of the visual field.

26. R. Desimone and J. Duncan, "Neural Mechanisms of Selective Visual Attention," *Annual Review of Neuroscience*, 18 (1995): 193–222; L. Chelazzi, E. K. Miller, J. Duncan, and R. Desimone, "Responses of Neurons in Macaque Area V4 during Memory-Guided Visual Search," *Cerebral Cortex*, 11 (2001): 761–772.

27. J. Driver and G. C. Baylis, "Attention and Visual Object Segmentation," in Parasuraman, *Attentive Brain*. Visual attention can directly modulate processing in primary visual areas such as V1 and the areas that process visual movement, such as MT and MST.

28. Z. W. Pylyshyn, "Visual Indexes, Preconceptual Objects, and Situated Vision," *Cognition*, 80 (2001): 159–177.

29. A. Treisman, "Feature Binding, Attention, and Object Perception," *Philosophical Transactions of the Royal Society B: Biological Sciences*, 353 (1998): 1295–1306.

30. It is clear that the parietal cortex is involved in attentional processes not only for vision but also in other modalities. Discrimination of sounds produces major activity in the inferior parietal cortex, the superior parietal cortex, and the inferior frontal cortex in conditions of dichotic listening. The inferior parietal and frontal cortex pair is activated in other attentional tasks of multimodal detection (visual-auditory, visual-tactile). In particular, the parietotemporal junction is activated during changes of stimuli, independent of their nature.

31. N. Kanwisher and E. Wojciulik, "Visual Attention: Insights from Brain Imaging," *Nature Reviews Neuroscience*, 1 (2000): 91–100. Other theories question too empirical a perspective and try to solve the problem of conflicts between early and late, and serial or parallel attention. Examples include the theories of Robert Desimone and John Duncan ("biased competition"), Anne Treisman ("integration of attributes"), and Nancy Kanwisher, who emphasizes the necessity of updating theories by combining top-down and bottom-up approaches and distinguishing objects and space. Jon Driver and Gordon C. Baylis criticize the metaphor of

Francis Crick's attentional spotlight in "Movement and Attention: The Spotlight Metaphor Breaks Down," *Journal of Experimental Psychology: Human Perception,* 15 (1989): 448–456.

32. Whalen and Phelps, *The Human Amygdala,* 362–381.

33. P. Mounoud, K. Duscherer, G. Moy, and S. Perraudin, "The Influence of Action Perception on Object Recognition: A Developmental Study," *Developmental Science,* 10 (2007): 836–852; J.-P. Gachoud, P. Mounoud, C. A. Hauert, and P. Viviani, "Motor Strategies in Lifting Movements: A Comparison of Adult and Child Performance," *Journal of Motor Behavior,* 15 (1983): 202–216.

34. A. Lubin, N. Poirel, S. Rossi, A. Pineau, and O. Houdé, "Math in Actions: Actor Mode Reveals the True Arithmetic Abilities of French-Speaking 2-Year-Olds in a Magic Task," *Journal of Experimental Child Psychology,* 103 (2009): 376–385; G. Leroux, J. Spiess, L. Zago, S. Rossi, A. Lubin, M. R. Turbelin, B. Mazoyer, N. Tzourio-Mazoyer, O. Houdé, and M. Joliot, "Adult Brains Don't Fully Overcome Biases That Lead to Incorrect Performance during Cognitive Development: An fMRI Study in Young Adults Completing a Piaget-Like Task," *Developmental Science,* 12 (2009): 326–338; O. Houdé and N. Tzourio-Mazoyer, "Neural Foundations of Logical and Mathematical Cognition," *Nature Reviews Neuroscience,* 4 (2003): 507–514; O. Houdé, L. Zago, E. Mellet, S. Moutier, A. Pineau, B. Mazoyer, and N. Tzourio-Mazoyer, "Shifting from the Perceptual Brain to the Logical Brain: The Neural Impact of Cognitive Inhibition Training," *Journal of Cognitive Neuroscience,* 12 (2000): 721–728.

35. A. Diamond, "All or None Hypothesis: A Global-Default Mode That Characterizes the Brain and Mind," *Development Psychology,* 45 (2009): 130–138; A. Diamond, "When in Competition against Engrained Habits, Is Conscious Representation Sufficient or Is Inhibition of the Habit Also Needed?" *Developmental Science,* 12 (2009): 20–22, discussion 24–25; A. Diamond, "Bootstrapping Conceptual Deduction Using Physical Connection: Rethinking Frontal Cortex," *Trends in Cognitive Science,* 10 (2006): 212–218; M. C. Davidson, D. Amso, L. C. Anderson, and A. Dia-

mond, "Development of Cognitive Control and Executive Functions from 4 to 13 Years: Evidence from Manipulations of Memory, Inhibition, and Task Switching," *Neuropsychologia*, 44 (2006): 2037–2078.

36. A. N. Meltzoff and M. K. Moore, "Newborn Infants Imitate Adult Facial Gestures," *Child Development*, 54 (1983): 702–709; A. N. Meltzoff and M. K. Moore, "Interpreting 'Imitative' Responses Early in Infancy," *Science*, 13 (1979): 217–219.

37. J.-D. Degos, A. C. Bachoud-Lévi, A. M. Ergis, C. Pétrissans, and P. Cesaro, "Selective Inability to Point to Extrapersonal Targets after Left Posterior Parietal Lesions," *Neurocase* 3 (1997): 31–39.

Chapter 5: The Brain as Emulator and Creator of Worlds

Epigraph. Personal communication, 2008. Jan Koenderink was among the first to suggest that the brain employs non-Euclidean geometries.

1. Details of this aspect of recent discoveries can be found in the major cognitive science journals, for example, *Trends in Cognitive Science*, *Trends in Neuroscience*, *Nature Reviews*, *Brain and Behaviour Science*, *Annals of Psychology*, and *Annals of Neuroscience*.

2. A. Michotte, "La Causalité physique est-elle une donnée phénoménale?" *Tijdschrift voor Philosophie*, 3 (1941): 290–328; A. Michotte, *The Perception of Causality*, trans. T. R. and E. Miles (Andover, MA: Methuen, 1962); A. Michotte et al., "Causalité, permanence et réalité phénoménale," *Studia Psychologica* (1962): 277–298.

3. Michotte, *Perception*, 21.

4. Michotte, *Perception*, 223.

5. J. McIntyre, M. Zago, A. Berthoz, and F. Lacquaniti, "Does the Brain Model Newton's Laws?" *Nature Neuroscience*, 4 (2001): 693–694.

6. R. A. Finke, J.-J. Freyd, and C. G. Shyi, "Implied Velocity and Acceleration Induce Transformations of Visual Memory," *Journal of Experimental Psychology: General*, 115 (1986): 175–188; R. A. Finke and J.-J. Freyd, "Transformations of Visual Memory Induced by Implied Motions of Pattern Elements," *Journal of Experimental Psychology: Learning, Memory, and Cognition*, 11 (1985): 780–794; V. S. Ramachandran and S. M. An-

stis, "Perceptual Organization in Moving Patterns," *Nature,* 304 (1983): 529–531.

7. H. Pieron cited in von Uexküll, *Umwelt und Innenwelt.*

8. J. J. Gibson, *The Senses Considered as Perceptual Systems* (London: George Allen and Unwin, 1968).

9. R. N. Shepard, *Mental Images and Their Transformations* (Cambridge, MA: MIT Press, 1982); R. N. Shepard, "Ecological Constraints on Internal Representations: Resonant Kinematics of Perceiving, Imaging, Thinking, and Dreaming," *Psychological Review,* 91 (1984): 417–447.

10. J. J. Koenderink, "'Controlled Hallucination' and 'Inverse Optics,'" *Perception,* 37 (2008): 87.

11. E. Mach, *The Science of Mechanics: A Critical and Historical Account of Its Development,* trans. T. McCormick (Chicago: Open Court Publishing, 1893).

12. G. Ganis, H. E. Schendan, and S. M. Kosslyn, "Neuroimaging Evidence for Object Model Verification Theory: Role of Prefrontal Control in Visual Object Categorization," *Neuroimage,* 34 (2007): 384–398.

13. Shepard, *Mental Images;* Shepard, "Ecological Constraints."

14. L. Cattaneo and G. Rizzolatti, "The Mirror Neuron System," *Archives of Neurology,* 66 (2009): 557–560; V. Caggiano, L. Fogassi, G. Rizzolatti, P. Thier, and A. Casile, "Mirror Neurons Differentially Encode the Peripersonal and Extrapersonal Space of Monkeys," *Science,* 324 (2009): 403–406; G. Rizzolatti, L. Fadiga, V. Gallese, and L. Fogassi, "Premotor Cortex and the Recognition of Motor Actions," *Brain Research: Cognitive Brain Research,* 3 (1996): 131–141; M. Gentilucci, L. Fogassi, G. Luppino, M. Matelli, R. Camarda, and G. Rizzolatti, "Somatotopic Representation in Inferior Area 6 of the Macaque Monkey," *Brain Behavior and Evolution,* 33 (1989): 118–121; G. Rizzolatti and C. Sinigaglia, *Les Neurones miroirs* (Paris: Odile Jacob, 2008).

15. H. Barlow, "Redundancy Reduction Revisited. Network: Computation in Neural Systems," *Neural Systems,* 12 (2001): 241–253.

16. Barlow, "Redundancy Reduction."

17. D. E. Koshland, "The Two-Component Pathway Comes to Eukaryotes," *Science,* 262 (1993): 532.

18. G. Horn, "Pathways of the Past: The Imprint of Memory," *Nature Reviews Neuroscience*, 5 (2004): 108–120.

19. K. D. Walton, D. Lieberman, A. Llinás, M. Begin, and R. R. Llinás, "Identification of a Critical Period for Motor Development in Neonatal Rats," *Neuroscience*, 51 (1992): 763–767.

20. I developed this idea in a book chapter in A. Berthoz, C. Ossola, and B. Stock, eds., "La Pluralité interprétative" (Paris: Collège de France, 2010). http://conferences-cdf.revues.org/147 (accessed 24 March 2011).

21. A German-Mongolian documentary film by Byambasuren Davaa and Luigi Falorni, *Die Geschichte vom weinenden Kamäl* (The story of the weeping camel), ARP Films, 2003.

22. For a recent review on these mechanisms, see F. Spiegel, "The Critical Period," *Current Biology*, 17 (2007): 742–743; N. Berardi, T. Pizzorusso, and L. Maffei, "Extracellular Matrix and Visualcortical Plasticity: Freeing," *Neuroscience*, 6 (2004): 877–888; S. B. Hofer, T. D. Mrsic-Flogel, T. Bonhoeffer, and M. Hübener, "Prior Experience Enhances Plasticity in Adult Visual Cortex," *Science*, 309 (2006): 2222–2226. B. D. Philpot, K. K. A. Cho, and M. F. Bear, "Obligatory Role of NR2A for Metaplasticity in Visual Cortex," *Neuron*, 53 (2007): 495–502.

23. Horn, "Pathways of the Past."

24. T. K. Hensch et al., "Local GABA Circuit Control of Experience-Dependent Plasticity in Local Cortical Circuits," *Nature Reviews Neuroscience*, 6 (2005): 877–888; S. Sugiyama, A. A. Di Nardo, S. Aizawa, I. Matsuo, M. Volovitch, A. Prochiantz, and T. K. Hensch, "Experience-Dependent Transfer of OTX2 Homeoprotein into the Visual Cortex Activates Postnatal Plasticity," *Cell*, 134 (2008): 508–520; T. Pizzorusso, P. Medini, N. Berardi, S. Chierzi, J. W. Fawcett, and L. Maffei, "Reactivation of Ocular Dominance Plasticity in the Adult Visual Cortex," *Science*, 298 (2002): 1248–1251. These efforts and others have provided the first proof that transmission involving GABA is necessary for plasticity in vivo. This confirms that the temporal evolution of the critical period can be controlled by inhibitor interneurons.

25. J. J. Gibson, *The Perception of the Visual World* (Boston: Houghton Mifflin, 1950); J. J. Gibson, *The Senses Considered as Perceptual Systems* (Bos-

ton: Houghton Mifflin, 1966); J. J. Gibson, "A Theory of Direct Visual Perception," in *The Psychology of Knowing*, ed. J. Royce and W. Rozenboom (New York: Gordon and Breach, 1972); J. J. Gibson, "The Theory of Affordances," in *Perceiving, Acting, and Knowing: Toward an Ecological Psychology*, ed. R. Shaw and J. Bransford (Hillsdale, NJ: Lawrence Erlbaum, 1977).

26. T. Kenet, D. Bibitchkov, M. Tsokyks, A. Grinvald, and A. Arieli, "Spontaneously Emerging Cortical Representations of Visual Attributes," *Nature*, 425 (2003): 954–956.

27. J. Droulez, personal communication.

Chapter 6: Simplexity in Perception

1. P. Bessière, C. Laugier, and R. Siegward, *Probabilistic Reasoning and Decision Making in Sensory-Motor Systems*, Springer Tracts in Advanced Robotics, vol. 46 (Heidelberg, Germany: Springer, 2008).

2. A. Triller and D. Choquet, "New Concepts in Synaptic Biology Derived from Single-Molecule Imaging," *Neuron*, 59 (2008): 359–374.

3. J.-B. Durand, R. Peeters, J.-F. Norman, J. T. Todd, and G. A. Orban, "Parietal Regions Processing Visual 3D Shape Extracted from Disparity," *Neuroimage*, 46 (2009): 1114–1126; S. S. Georgieva, J. T. Todd, R. Peeters, and G. A. Orban, "The Extraction of 3D Shape from Texture and Shading in the Human Brain," *Cerebral Cortex*, 18 (2008): 2416–2438; G. A. Orban, "Higher Order Visual Processing in Macaque Extrastriate Cortex," *Physiological Review*, 88 (2008): 59–89.

4. F. Klam, J. Petit, A. Grantyn, and A. Berthoz, "Predictive Elements in Ocular Interception and Tracking of a Moving Target by Untrained Cats," *Experimental Brain Research*, 139 (2001): 233–234; E. Olivier, A. Grantyn, M. Chat, and A. Berthoz, "The Control of Slow Orienting Eye Movements by Tectoreticulospinal Neurons in the Cat: Behavior, Discharge Patterns, and Underlying Connections," *Experimental Brain Research*, 93 (1993): 35–49; A. Grantyn and A. Berthoz, "Reticulo-Spinal Neurons Participating in the Control of Synergic Eye and Head Movements during Orienting in the Cat. I. Behavioral Properties," *Experimental Brain Research*, 66 (1987): 339–354.

5. L. Petit, C. Orssaud, N. Tzourio, F. Grivello, A. Berthoz, and B. Mazoyer, "PET Study of Voluntary Saccadic Eye Movements in Humans," *Journal of Neurophysiology*, 69 (1993): 1009–1017.

6. M.-H. Grosbras and A. Berthoz, "Parieto-frontal Networks and Gaze Shifts in Humans: Review of Functional Magnetic Resonance Imaging Data," *Advances in Neurology*, 93 (2003): 269–280; M.-H. Grosbras, E. Lobel, P.-F. Van de Moortele, D. LeBihan, and A. Berthoz, "An Anatomical Landmark for the Supplementary Eye Fields in Human Revealed with Functional Magnetic Resonance Imaging," *Cerebral Cortex*, 9 (1999): 705–711.

7. On extrastriate body area, see M. Spiridon, B. Fischl, and N. Kanwisher, "Location and Spatial Profile of Category-Specific Regions in Human Extrastriate Cortex," *Human Brain Mapping*, 27 (2006): 77–89; C. Urgesi, B. Calvo-Merino, P. Haggard, and S. M. Aglioti, "Transcranial Magnetic Stimulation Reveals Two Cortical Pathways for Visual Body Processing," *Journal of Neuroscience*, 27 (2007): 8023–8030.

8. R. Kiani, H. Esteky, K. Mirpour, and K. Tanaka, "Object Category Structure in Response Patterns of Neuronal Population in Monkey Inferior Temporal Cortex," *Journal of Neurophysiology*, 97 (2007): 4296–4309; S. R. Lehky and K. Tanaka, "Enhancement of Object Representations in Primate Perirhinal Cortex during a Visual Working Memory Task," *Journal of Neurophysiology*, 97 (2007): 1298–1310; W. Suzuki, K. Matsumoto, and K. Tanaka, "Neuronal Responses to Object Images in the Macaque Inferotemporal Cortex at Different Stimulus Discrimination Levels," *Journal of Neuroscience*, 41 (2006): 10524–10535.

9. H. J. Spiers and E. A. Maguire, "The Dynamic Nature of Cognition during Wayfinding," *Journal of Environmental Psychology*, 28 (2008): 232–249; J. J. Summerfield, D. Hassabis, and E. A. Maguire, "Cortical Midline Involvement in Autobiographical Memory," *Neuroimage*, 44 (2009): 1188–1200; K. Woollett, J. Glensman, and E. A. Maguire, "Non-Spatial Expertise and Hippocampal Gray Matter Volume in Humans," *Hippocampus*, 18 (2008): 981–984; C. I. Baker, T. L. Hutchison, and N. Kanwisher, "Does the Fusiform Face Area Contain Subregions Highly Selective for Nonfaces?" *Nature Neuroscience*, 10 (2007): 3–4; F. Vinckier,

S. Dehaene, A. Jobert, J.-P. Dubus, M. Sigman, and L. Cohen, "Hierarchical Coding of Letter Strings in the Ventral Stream: Dissecting the Inner Organization of the Visual Word-Form System," *Neuron*, 55 (2007): 143–156.

10. E. Koechlin and A. Hyafil, "Anterior Prefrontal Function and the Limits of Human Decision-Making," *Science*, 318 (2007): 594–598.

11. The technical explanation of the process that permits this linearization is the following: Dense firing produces a divisive shunting inhibition that normalizes and linearizes the cortical response. One can imagine that the feedback of the higher areas prepares the integrating operations of the primary cortex by positioning the network at a high level, which linearizes the sensory responses, shortens their latency, and consequently increases their speed.

12. D. N. Lee, "A Theory of Visual Control of Braking Based on Information about Time-to-Collision," *Perception*, 5 (1976): 437–459.

13. See n. 4 of this chapter and A. Grantyn, V. Ong-Mean Jacques, and A. Berthoz, "Reticulo-Spinal Neurons Participating in the Control of Synergic Eye and Head Movements during Orienting in the Cat. II. Morphological Properties as Revealed by Intra-Axonal Injections of Horseradish Peroxidase," *Experimental Brain Research*, 66 (1987): 355–377; A. Berthoz, A. Grantyn, and J. Droulez, "Some Collicular Efferent Neurons Code Saccadic Eye Velocity," *Neuroscience Letters*, 72 (1986): 289–294.

14. G. A. Calvert, C. Spence, and B. E. Stein, *The Handbook of Multisensory Processes* (Cambridge, MA: MIT Press, 2004), 343–356.

15. M. Poincaré, *Science and Hypothesis*, trans. G. B. Halsted (New York: Science Press, 1905).

16. R. Llinás, *I of the Vortex*.

17. K. O. Johnson, "The Roles and Functions of Cutaneous Mechanoreceptors," *Current Opinion in Neurobiology*, 11 (2001): 455–461.

18. The corollary discharge is a copy addressed to the perceptual centers of the motor command. I discussed the importance of this signal and its neural basis in *The Brain's Sense of Movement*.

19. P. Guillaume, *La Psychologie de la forme* (Paris: Flammarion, 1937).

20. J. von Uexküll, "An Introduction to *Umwelt*," *Semiotica*, 134 (1936/2001): 107–110.

21. J. Ruben, J. Schwiemann, M. Deuchert, R. Meyer, T. Krause, G. Curio, K. Villringer, R. Kurth, and A. Villringer, "Somatotopic Organization of Human Secondary Somatosensory Cortex," *Cerebral Cortex*, 11 (2001): 463–473.

22. C. Keysers, B. Wicker, V. Gazzola, J. L. Angon, L. Fogassi, and V. Gallese, "A Touching Sight: SII/PV Activation during the Observation and Experience of Touch," *Neuron*, 42 (2004): 335–346.

23. D. N. Saito, T. Okada, Y. Morita, Y. Yonekura, and N. Sadato, "Tactile-Visual Cross-Modal Shape Matching: A Functional MRI Study," *Brain Research: Cognitive Brain Research*, 17 (2003): 14–25.

24. A. Lécuyer, M. Vidal, O. Joly, C. Mégard, and A. Berthoz, "Can Haptic Feedback Improve the Perception of Self-Motion in Virtual Reality?" (Paper presented at Haptics Symposium, Chicago, March 27–28, 2004).

25. Y. Komura, R. Tamura, T. Uwano, H. Nishijo, K. Kaga, and T. Ono, "Retrospective and Prospective Coding for Predicted Reward in the Sensory Thalamus," *Nature*, 412 (2001): 546–549.

26. My thanks to Vincent Hayward, who helped me to edit what follows on the haptic sense. Our discussions were lively, owing to his immense theoretical knowledge and the keenness of his experimental intuition. I cite only a few of his contributions.

27. R. S. Johansson and J. R. Flanagan, "Coding and Use of Tactile Signals from the Fingertips in Object Manipulation Tasks," *Nature Reviews Neuroscience*, 10 (2009): 345–359; J. R. Flanagan, M. C. Bowman, and R. S. Johansson, "Control Strategies in Object Manipulation Tasks," *Current Opinion in Neurobiology*, 16 (2009): 650–659.

28. B. B. Edin and N. Johnsson, "Skin Strain Patterns Provide Kinesthetic Information to the Human Central Nervous System," *Journal of Physiology*, 487 (1995): 243–251; G. Robles De La Torre and V. Hayward, "Force Can Overcome Object Geometry in the Perception of Shape through Active Touch," *Nature*, 412 (2001): 445–448; H. Dosmohamed

and V. Hayward, "Trajectory of Contact Region on the Fingerpad Gives the Illusion of Haptic Shape," *Experimental Brain Research*, 164 (2005): 387–394.

29. Hertz (Hz) is a dimension that denotes the inverse of time or cycles per second of an oscillatory process. For example, the A (above middle C) of a diapason (a tool used by musicians to calibrate sounds) corresponds to a metal vibration of 440 times per second, or 440 Hz. A single vibration lasts a little more than 2 milliseconds. The oscillations of the neural networks in mammals, for example, can go from 0.1 Hz (or even less) in the cortex of a sleeping mammal to 200 Hz or more in certain highly specialized structures.

30. Generally speaking, the activity of a neuron is measured in number of action potentials (APs) per second, that is, also in Hertz. To do it, one divides time up into short intervals, counts the number of APs emitted in each interval, and averages them over the sum of all trials.

31. Each odor is symbolized by a category that occupies a certain volume in the space of these categories. One category is the collection of nested categories. Conversely, the precise identity of one category corresponds to the difference between that category and its neighbors (the volume of space of that category overlaps with no other).

32. M. Rabinovich, A. Volkovskii, P. Lecanda, R. Huerta, H. D. I. Abarbanel, and G. Laurent, "Dynamical Encoding by Networks of Competing Neuron Groups: Winnerless Competition," *Physical Review Letters*, 87 (2001): 068102.

33. A. Peyrache, M. Khamassi, K. Benchenane, S. I. Wiener, and F. P. Battaglia, "Replay of Rule-Learning Related Neural Patterns in the Prefrontal Cortex during Sleep," *Nature Neuroscience*, 12 (2009): 919–926.

Chapter 7: The Laws of Natural Movement

1. P. Viviani and N. Stucchi, "Biological Movements Look Uniform: Evidence for Motor-Perceptual Interactions," *Journal of Experimental Psychology (Human Perception)*, 18 (1992): 603–623.

2. G. Cappellini, Y. P. Ivanenko, R. E. Poppele, and F. J. Lacquaniti, "Motor

Patterns in Human Walking and Running," *Journal of Neurophysiology*, 95 (2006): 3426–3437. See also Y. P. Ivanenko et al., "Modular Control of Limb Movements during Human Locomotion," *Journal of Neuroscience*, 27 (2007): 11149–11161.

3. B. Calvo-Merino, J. Grèzes, D. E. Glaser, R. E. Passingham, and P. Haggard, "Seeing or Doing? Influence of Visual and Motor Familiarity in Action Observation," *Current Biology*, 16 (2006): 1905–1910; erratum, 16 (2006): 2277.

4. E. Todorov and Z. Ghahramani, "Unsupervised Learning of Sensory-Motor Primitives," *Proceedings of the Annual International Conference of the IEEE Engineering in Medicine and Biology Society*, 5 (2003): 1750–1753.

5. I described this organization in *The Brain's Sense of Movement* and *Emotion and Reason*.

6. M. Jeannerod, "The Formation of Finger Grip During Prehension. A Cortically Mediated Visuomotor Pattern," *Behavioural and Brain Research*, 19 (1986): 99–116.

7. McIntyre et al., "Does the Brain Model Newton's Laws?"; I. Indovina, V. Maffei, G. Bosco, M. Zago, E. Macaluso, and F. Lacquaniti, "Representation of Visual Gravitational Motion in the Human Vestibular Cortex," *Science*, 308 (2005): 416–419.

8. I briefly touched on this problem in *Emotion and Reason*.

9. D. W. Franklin, E. Burdet, K. P. Tee, R. Osu, C. M. Chew, T. E. Milner, and M. Kawato, "CNS Learns Stable, Accurate, and Efficient Movements using a Simple Algorithm," *Journal of Neuroscience*, 28 (2008): 11165–11173; E. Burdet, D. W. Franklin, R. Osu, K. P. Tee, M. Kawato, and T. E. Milner, "How Are Internal Models of Unstable Tasks Formed?" *Proceedings of the Annual International Conference of the IEEE Engineering in Medicine and Biology Society*, 6 (2004): 4491–4494; N. Schweighofer, J. Spoelstra, M. A. Arbib, and M. Kawato, "Role of the Cerebellum in Reaching Movements in Humans. II. A Neural Model of the Intermediate Cerebellum," *European Journal of Neuroscience*, 10 (1998): 95–105; Hanneton et al., "Does the Brain Use Sliding Variables?"

10. H. Imamizu, S. Miyauchi, T. Tamada, Y. Sasaki, R. Takino, B. Pütz, T. Yoshioka, and M. Kawato, "Human Cerebellar Activity Reflecting an Acquired Internal Model of a New Tool," *Nature*, 403 (2000): 192–195.

11. S. Pasalar, A. V. Roitman, W. K. Durfee, and T. J. Ebner, "Force Field Effects on Cerebellar Purkinje Cell Discharge with Implications for Internal Models," *Nature Neuroscience*, 9 (2006): 1404–1411.

12. S. G. Massaquoi and J.-J. Slotine, "The Intermediate Cerebellum May Function as a Wave-Variable Processor," *Neuroscience Letters*, 215 (1996): 60–64; J. McIntyre and J.-J. Slotine, "Does the Brain Make Waves to Improve Stability?" *Brain Research Bulletin*, 75 (1996): 712–722.

13. Hanneton et al., "Does the Brain Use Sliding Variables?"; F. Klam and W. Graf, "Vestibular Response Kinematics in Posterior Parietal Cortex Neurons of Macaque Monkeys," *European Journal of Neuroscience*, 18 (2003): 995–1010; F. Klam, J. Petit, A. Grantyn, and A. Berthoz, "Predictive Elements in Ocular Interception and Tracking of a Moving Target by Untrained Cats," *Experimental Brain Research*, 139 (2001): 233–247.

14. Models of this sort have been proposed by Alexandre Pouget: S. Denève, J. R. Duhamel, and A. Pouget, "Optimal Sensorimotor Integration in Recurrent Cortical Networks: A Neural Implementation of Kalman Filters," *Journal of Neuroscience*, 27 (2007): 5744–5756; A. Pouget, S. Denève, and J.-R. Duhamel, "A Computational Perspective on the Neural Basis of Multisensory Spatial Representations," *Nature Reviews Neuroscience*, 3 (2002): 741–747.

15. R. A. Rescorla and A. R. Wagner, "A Theory of Pavlovian Conditioning: Variations in the Effectiveness of Reinforcement and Nonreinforcement," in *Classical Conditioning II*, ed. A. H. Black and W. R. Prokasy (New York: Appleton-Century-Crofts, 1972), 64–99.

16. M. O. Ernst and M. S. Banks, "Humans Integrate Visual and Haptic Information in a Statistically Optimal Fashion," *Nature*, 415 (2002): 429–433; J. Laurens and J. Droulez, "Bayesian Processing of Vestibular Information," *Biological Cybernetics*, 96 (2007): 389–404.

17. H. Barlow, "Redundancy Reduction."

18. M. N. Shadlen and W. T. Newsome, "Neural Basis of a Perceptual Deci-

sion in the Parietal Cortex (Area LIP) of the Rhesus Monkey," *Journal of Neurophysiology*, 86 (2001): 1916–1936; A. N. McCoy and M. L. Platt, "Expectations and Outcomes: Decision-Making in the Primate Brain," *Journal of Comparative Physiology A: Neuroethology, Sensory, Neural, and Behavioral Physiology*, 191 (2005): 201–211.

19. S. Denève, P.-E. Latham, and A. Pouget, "Efficient Computation and Cue Integration with Noisy Population Codes," *Nature Neuroscience*, 4 (2001): 826–831; S. Denève, P.-E. Latham, and A. Pouget, "Reading Population Codes: A Neural Implementation of Ideal Observers," *Nature Neuroscience*, 2 (1999): 740–745.

20. A Kalman filter is an operator that predicts and optimizes processing of information. It is widely used in robotics and has also been employed in modeling sensory processing.

21. I cite here the comments of the mathematician and modeler Oliver Faugeras, whom I interviewed regarding simplicity: "The problem is how to define what we mean by 'simplify.' This assumes that it is possible to quantify the complexity of a calculation, and that is where things get complicated. As far as I know, Kolmogorov is the only person to have attempted to respond to this question in a formal way, and the result was the concept of 'Kolmogorov complexity.' Briefly, the crux is the smallest program that is capable of carrying out a given calculation. The problem is that no existing algorithm can calculate this size automatically. Moreover, the notion of 'simple calculation' also depends on the material you have. A line of code is sufficient to program an integrator on a computer. You might think that it is enough to ask what is simple for a neuron, but there again, the answer is far from clear. With all the ionic channels and neurotransmitters known, you can carry out nearly any calculation. Tony Bell, Larry Abbott, Bartlet Mel, and others since them have shown that." But it is much more complicated with neurons, for example, Jacques Droulez's dynamic memory network. [Cf. J. Droulez and A. Berthoz, "A Neural Network Model of Sensorimotor Maps with Predictive Short-Term Memory Properties," *Proceedings of the National Academy of Sciences of the United States of America*, 88 (1991): 9653–9657.]

22. See G. Edelman and G. Tononi, *A Universe of Consciousness: How Matter Becomes Imagination* (New York: Basic Books, 2000).

23. J.-J. Slotine and W. Lohmiller, "Modularity, Evolution, and the Binding Problem: A View from Stability Theory," *Neural Networks*, 14 (2001): 137–145.

24. W. Wang and J.-J. Slotine, "On Partial Contraction Analysis for Coupled Nonlinear Oscillators," *Biological Cybernetics*, 92 (2005): 38–53.

25. B. Girard, N. Tabareau, J.-J. Slotine, and A. Berthoz, "Contracting Model of the Basal Ganglia," in *Modelling Natural Action Selection: Proceedings of an International Workshop*, ed. J. Bryson, T. Prescott, and A. Seth (Brighton, UK: AISB Press, 2005), 69–76; B. Girard, N. Tabareau, A. Berthoz, and J.-J. Slotine, "Selective Amplification using a Contracting Model of the Basal Ganglia," in *NeuroComp 2006*, ed. F. Alexandre, Y. Boniface, L. Bougrain, B. Girau, and N. Rougier, 30–33.

Chapter 8: The Simplex Gesture

1. I described this proposal with J. L. Petit in *Phénoménologie et physiologie de l'action* (Paris: Odile Jacob, 2008).

2. J. Baltrušaitis, *Le Moyen-Âge fantastique. Antiquités et exotismes dans l'art gothique* (Paris: Flammarion, 2008).

3. J.-C. Schmitt, *La Raison des gestes dans l'Occident médiéval* (Paris: Gallimard, 1990).

4. The notion of *habitus* was used by Pierre Bourdieu.

5. V. Meyerhold, *Meyerhold on Theatre*, vol. 2, 1917–1930, trans. and ed. E. Braun (London: Methuen, 1969).

6. Sara Longo has written an unpublished paper on this subject.

7. B. Pasquinelli, *Le Geste et l'expression*, trans. C. Mulkai (Paris: Hazan, 2006).

8. L. Reau, *Iconographie de l'art chrétien* (Paris: PUF, 1955–1959).

9. Pasquinelli, *Le Geste*.

10. A. de Libera, *La Quête de l'identité*, Archéologie du sujet II (Paris: Vrin, 2008); A. de Libera, *Naissance du sujet*, Archéologie du sujet I (Paris: Vrin, 2007).

11. Cicero, *On the Ideal Orator*. In this long extract from *On the Orator*, an epistolary treatise composed at the request of the celebrated Brutus in A.D. 46, Cicero highlights the different kinds of eloquence through the person of an ideal orator.

12. E. G. Slingerland, *Effortless Action: Wu-Wei as Conceptual Metaphor and Spiritual Ideal in Early China* (New York: Oxford University Press, 2003).

13. T. Ribot, *La Psychologie des sentiments* (Paris: L'Harmattan, 1896).

14. R. Recht, *Le Croire et le voir. L'Art des cathédrales (XII–XV siècles)* (Paris: Gallimard, 1999).

15. Schmitt, *La Raison des gestes;* J.-C. Schmitt, "Gestus-Gesticulatio. Contribution à l'étude du vocabulaire latin médiéval des gestes," in *La Lexicographie du latin médiéval et ses rapports avec les recherches actuelles sur la civilisation du Moyen Age* (Paris: CNRS, 1981), 377–390.

16. P. Ricoeur, *Soi-même comme un autre* (Paris: Seuil, 1990).

17. C. Darwin, *The Expression of the Emotions in Man and Animals* (Chicago: University of Chicago Press, 1965).

18. J. Panksepp, *Affective Neuroscience: The Foundations of Human and Animal Emotions* (New York: Oxford University Press, 1998).

19. D. C. Blanchard, A. L. Hynd, K. A. Minke, T. Minemoto, and R. J. Blanchard, "Human Defensive Behaviors to Threat Scenarios Show Parallels to Fear and Anxiety-Related Defense Patterns of Non-Human Mammals," *Neuroscience and Biobehavioral Reviews*, 25 (2001): 761–770.

20. S. Pichon, B. de Gelder, and J. Grèzes, "Two Different Faces of Threat. Comparing the Neural Systems for Recognizing Fear and Anger in Dynamic Body Expressions," *Neuroimage*, 14 (2009); J. Grèzes, S. Pichon, and B. de Gelder, "Perceiving Fear in Dynamic Body Expressions," *Neuroimage*, 35 (2007): 959–967; B. de Gelder, "Towards the Neurobiology of Emotional Body Language," *Nature Reviews Neuroscience*, 7 (2006): 242–249.

21. M. S. Graziano, C. S. Taylor, and T. Moore, "Probing Cortical Function with Electrical Stimulation," *Nature Neuroscience*, 5 (2002): 921.

Chapter 9: Walking: A Challenge to Complexity

Epigraph. Gell-Mann, *The Quark and the Jaguar*, 99–100.

1. A. Berthoz and J.-L. Petit, *Physiologie de l'action et phénoménologie* (Paris: Odile Jacob, 2006).

2. T. Mentel, L. Cangiano, S. Grillner, and A. Büschges, "Neuronal Substrates for State-Dependent Changes in Coordination between Motoneuron Pools during Fictive Locomotion in the Lamprey Spinal Cord," *Journal of Neuroscience*, 28 (2008): 868–879; S. Grillner, A. Kozlov, P. Dario, C. Stefanini, A. Menciassi, A. Lansner, and J. Hellgren Kotaleski, "Modeling a Vertebrate Motor System: Pattern Generation, Steering, and Control of Body Orientation," *Progress in Brain Research*, 165 (2007): 221–234; S. Grillner, P. Wallén, K. Saitoh, A. Kozlov, and B. Robertson, "Neural Bases of Goal-Directed Locomotion in Vertebrates—An Overview," *Brain Research Review*, 57 (2008): 2–12.

3. A. Ijspeert, A. Crespi, D. Ryczko, and J.-M. Cabelguen, "From Swimming to Walking with a Salamander Robot Driven by a Spinal Cord Model," *Science*, 315 (2007): 1416–1420.

4. J. Soechting and M. Flanders, "Moving in Three Dimensional Space: Frames of Reference, Vectors, and Coordinate Systems," *Annual Review of Neuroscience*, 15 (1992): 167–191.

5. Y. P. Ivanenko, G. Cappellini, N. Dominici, R. E. Poppele, and F. Lacquaniti, "Modular Control of Limb Movements during Human Locomotion," *Journal of Neuroscience*, 27 (2007): 11149–11161; G. Cappellini, Y. P. Ivanenko, R. E. Poppele, and F. Lacquaniti, "Motor Patterns in Human Walking and Running," *Journal of Neurophysiology*, 95 (2006): 3426–3437; H. Hicheur, A. V. Terekhov, and A. Berthoz, "Intersegmental Coordination during Human Locomotion: Does Planar Covariation of Elevation Angles Reflect Central Constraints?" *Journal of Neurophysiology*, 96 (2006): 1406–1419; A. Barliya, L. Omlor, M. A. Giese, and T. Flash, "An Analytical Formulation of the Law of Intersegmental Coordination during Human Locomotion," *Experimental Brain Research*, 193 (2009): 371–385.

6. Y. P. Ivanenko, A. d'Avella, R. E. Poppele, and F. Lacquaniti, "On the

Origin of Planar Covariation of Elevation Angles during Human Locomotion," *Journal of Neurophysiology*, 99 (2008): 1890–1898.

7. Gell-Mann, *The Quark and the Jaguar*.

8. H. Geyer, "Simple Models of Legged Locomotion Based on Compliant Limb Behavior" (PhD diss., Friedrich Schiller University, 2005); H. Geyer, A. Seyfarth, and R. Blickhan, "Compliant Leg Behavior Explains Basic Dynamics of Walking and Running," *Proceedings of Biological Science*, 273 (2006): 2861–2867; S. Collins, A. Ruina, R. Tedrake, and M. Wisse, "Efficient Bipedal Robots Based on Passive-Dynamic Walkers," *Science*, 307 (2005): 1082–1085, and *Science*, 308 (2005): 58–59.

9. E. Burdet, K. P. Tee, I. Mareels, T. E. Milner, C. M. Chew, D. W. Franklin, R. Osu, and M. Kawato, "Stability and Motor Adaptation in Human Arm Movements," *Biological Cybernetics*, 94 (2006): 20–32.

10. S. Vernazza-Martin et al., "Kinematic Synergy Adaptation to an Unstable Support Surface and Equilibrium Maintenance during Forward Trunk Movement," *Experimental Brain Research*, 173 (2006): 62–78.

11. L. Nashner, "Adapting Reflexes Controlling the Human Posture," *Experimental Brain Research*, 26 (1976): 59–72; L. Nashner and A. Berthoz, "Visual Contribution to Rapid Motor Responses during Postural Control," *Brain Research*, 150 (1978): 403–407.

12. On the question of the organizational flexibility of complex movements, see P. J. Cordo and V. S. Gurfintrel, "Motor Coordination Can Be Fully Understood Only by Studying Complex Movements," *Progress in Brain Research*, 143 (2004): 29–38; P. J. Cordo and L. M. Nashner, "Properties of Postural Adjustments Associated with Rapid Arm Movements," *Journal of Neurophysiology*, 47 (2004): 287–302.

13. A. Berthoz, "Reference Frames for the Perception and Control of Movement," in *Brain and Space*, ed. J. Paillard (Oxford: Oxford University Press, 1991), 82–111; see also the work of F. Lacquaniti and J. McIntyre.

14. T. Pozzo, A. Berthoz, and L. Lefort, "Head Stabilization during Various Locomotor Tasks in Humans. I. Normal Subjects," *Experimental Brain Research*, 82 (1990): 97–1016. On the flexibility of reference frames in postural control, see also T. Pozzo, G. Clément, and A. Berthoz, "Motor

Control of a Handstand," *Agressologie*, 29 (1988): 649–651; G. Clément, T. Pozzo, and A. Berthoz, "Contribution of Eye Positioning to Control the Upside-Down Standing Posture," *Experimental Brain Research*, 73 (1988): 569–576; M. A. Hollands, K. L. Sorensen, and A. E. Patla, "Effects of Head Immobilization on the Coordination and Control of Head and Body Reorientation and Translation during Steering," *Experimental Brain Research*, 140 (2001): 223–233.

15. R. Grasso, C. Assaiante, P. Prévost, and A. Berthoz, "Development of Anticipatory Orienting Strategies during Locomotor Tasks in Children," *Neuroscience and Biobehavioral Review*, 22 (1998): 533–539.

16. E. Lobel, J.-F. Kleine, D. L. Bihan, A. Leroy-Willig, and A. Berthoz, "Functional MRI of Galvanic Vestibular Stimulation," *Journal of Neurophysiology*, 80 (1998): 269–270.

17. P. Kahane, D. Hoffmann, L. Minotti, and A. Berthoz, "Reappraisal of the Human Vestibular Cortex by Cortical Electrical Stimulation Study," *Annals of Neurology*, 54 (2003): 615–624.

18. W. Penfield, *The Mystery of the Mind: A Critical Study of Consciousness and the Human Brain* (Princeton, NJ: Princeton University Press, 1975); W. Penfield, "Vestibular Sensation," *Comptes rendus, 4ème Congrès neurologique international*, vol. 3 (Paris: Masson, 1949).

19. This idea is currently the object of a thesis by Romain David in collaboration with the team of Profs. Taquet and Janvier of the National Museum of Natural History in Paris.

20. A. Corbin, J.-J. Courtine, and G. Vigarello, eds., *Histoire du corps. III. Les Mutations du regard. Le 20ème siècle* (Paris: Seuil, 2006).

21. P. P. Vidal, C. de Waele, W. Graf, and A. Berthoz, "Skeletal Geometry underlying Head Movements," *Annals of the New York Academy of Science*, 545 (1988): 228–238; P. P. Vida, W. Graf, and A. Berthoz, "The Orientation of the Cervical Vertebral Column in Unrestrained Awake Animals. I. Resting Position," *Experimental Brain Research*, 61 (1986): 549–559. A. Berthoz, W. Graf, and P. P. Vidal, *The Head-Neck Sensory Motor System* (Oxford: Oxford University Press, 1991).

22. H. Hicheur, S. Vieilledent, M. J. E. Richardson, T. Flash, and A. Berthoz,

"Velocity and Curvature in Human Locomotion along Complex Curved Paths: A Comparison with Hand Movements," *Experimental Brain Research*, 162 (2005): 145–154.

23. F. Polyakov, E. Start, R. Drori, M. Abeles, and T. Flash, "Parabolic Primitives and Cortical States: Merging Optimality with Geometric Invariances," *Biological Cybernetics*, 100 (2009): 159–184.

24. D. Bennequin, R. Rusch, A. Berthoz, and T. Flash, "Movement Timing and Invariance Arise from Several Geometrics," *PLoS Computational Biology*, 5 (2009): e1000426.

25. F. Lacquaniti, C. Terzuolo, and P. Viviani, "The Law Relating the Kinematic and Figural Aspects of Drawing Movements," *Acta Psychologica*, 54 (1983): 115–130; P. Viviani and C. Terzuolo, "Trajectory Determines Movement Dynamics," *Neuroscience*, 7 (1982): 431–437.

26. P. Viviani made the initial discovery of the relationship between perception and action using this law, as I mentioned in *The Brain's Sense of Movement;* P. Viviani and T. Flash, "Minimum-Jerk, Two-Thirds Power Law, and Isochrony: Converging Approaches to Movement Planning," *Journal of Experimental Psychology: Human Perception and Performance*, 21 (1995): 32–53; E. Dayan, A. Casile, N. Levit-Biknnun, M. A. Giese, T. Hendler, and T. Flash, "Neural Representations of Kinematic Laws of Motion: Evidence for Action-Perception Coupling," *Proceedings of the National Academy of Sciences of the United States of America*, 104 (2007): 20582–20587.

27. U. Maoz, A. Berthoz, and T. Flash, "Complex Unconstrained Three-Dimensional Hand Movement and Constant Equi-Affine Speed," *Journal of Neurophysiology*, 101 (2009): 1002–1015.

28. S. Vieilledent, S. Dalbera, Y. Kerlirzin, and A. Berthoz, "Relationship between Velocity and Curvature of a Human Locomotor Trajectory," *Neuroscience Letters*, 305 (2001): 65–69.

29. Hicheur et al., "Velocity and Curvature in Human Locomotion."

30. Bennequin et al., "Movement Timing and Invariance." Faugeras has suggested the use of non-Euclidean geometries for vision. O. Faugeras, "Cartan's Moving Frame Method and Its Application to the Geometry

and Evolution of Curves in the Euclidian Affine and Projective Planes," in *Applications of Invariants in Computer Vision*, ed. J. L. Mundy (London: Springer, 2009), 11–46.

Chapter 10: Simplex Space

Epigraph. James, *Textbook of Psychology, 294.*

1. A. Prochiantz, *Les Anatomies de la pensée. A quoi pensent les calamars?* (Paris: Odile Jacob, 1997).

2. P. Dusapin, *Une musique en train de se faire* (Paris: Seuil, 2009). One can see in this book the use that composers make of space.

3. E. Bartfeld and A. Brinvald, "Relationships between Orientation-Preference Pinwheels, Cytochrome Oxidase Blobs, and Ocular-Dominance Columns in Primate Striate Cortex," *Proceedings of the National Academy of Sciences of the United States of America*, 89 (1992): 11905–11909; H. Y. Lee, M. Yahyanejad, and M. Kardar, "Symmetry Considerations and Development of Pinwheels in Visual Maps," *Proceedings of the National Academy of Sciences of the United States of America*, 100 (2003): 16036–16040; K. Ohki, S. Chung, P. Kara, M. Hubener, T. Bonhoeffer, and R. C. Reid, "Highly Ordered Arrangement of Single Neurons in Orientation Pinwheels," *Nature*, 442 (2006): 925–928.

4. J. Petitot, "The Neurogeometry of Pinwheels as a sub-Riemannian Contact Structure," *Journal of Physiology*, 97 (2003): 265–309.

5. D. Boussaoud and F. Bremmer, "Gaze Effects in the Cerebral Cortex: Reference Frames for Space Coding and Action," *Experimental Brain Research*, 128 (1999): 170–180.

6. In the rat, retinotopy is present in the colliculus but not the logarithmic expansion; its map is linear. This logarithmic distribution of the map is not found only in the colliculus.

7. N. Tabareau, D. Bennequin, A. Berthoz, J.-J. Slotine, and B. Girard, "Geometry of the Superior Colliculus Mapping and Efficient Oculomotor Computation," *Biological Cybernetics*, 97 (2007): 279–292.

8. P. H. Weiss, J. C. Marshall, G. Wunderlich, L. Tellmann, P. W. Halligan, H. J. Freund, K. Zilles, and G. R. Fink, "Neural Consequences of Acting

in Near versus Far Space: A Physiological Basis for Clinical Dissociations," *Brain*, 123 (2000): 2531–2541.

9. Bennequin et al., "Movement Timing and Invariance."

10. T. Flash and A. A. Handzel, "Affine Differential Geometry Analysis of Human Arm Movements," *Biological Cybernetics* 96 (2007): 577–601.

11. See H. Weil on symmetries. See also F. Bailly and G. Longo, *Mathématiques et sciences de la nature. La Singularité physique du vivant* (Paris: Hermann, 2006).

12. A. Berthoz and G. Jorland, *L'Empathie* (Paris: Odile Jacob, 2004); B. Thirioux, G. Jorland, M. Bret, M.-H. Tramus, and A. Berthoz, "Walking on a Line: A Motor Paradigm using Rotation and Reflection Symmetry to Study Mental Body Transformations," *Brain and Cognition*, 70 (2004): 191–200.

13. S. Lambrey et al., "Distinct Visual Perspective-Taking Strategies Involve the Left and Right Medial Temporal Lobe Structures Differently," *Brain*, 131 (2008): 523–534.

14. D. Kimura, *Sex and Cognition* (Cambridge, MA: MIT Press, 1999); M. Hines, *Psychoanalysis and Neuroscience: Brain Gender* (New York: Oxford University Press, 2004); S. Lambrey and A. Berthoz, "Gender Differences in the Use of External Landmarks versus Spatial Representations Updated by Self-Motion," *Journal of Integrative Neuroscience*, 6 (2007): 379–401; L. Cahill, "Why Sex Matters for Neuroscience," *Nature Reviews Neuroscience*, 7 (2006): 477–484.

15. O. Ghaëm, E. Mellet, F. Crivello, N. Tzourio, B. Mazoyer, A. Berthoz, and M. Denis, "Mental Navigation along Memorized Routes Activates the Hippocampus, Precuneus, and Insula," *NeuroReport*, 8 (1997): 739–744. G. Vallar, E. Lobel, G. Galati, A. Berthoz, L. Pizzamiglio, and D. Le Bihan, "The Neural Basis of Egocentric and Allocentric Coding of Space in Humans: A Functional Magnetic Resonance Study," *Experimental Brain Research*, 133 (1999): 156–164; E. Mellet, S. Briscogne, N. Tzourio-Mazoyer, O. Ghaëm, L. Petit, L. Zago, O. Etard, A. Berthoz, B. Mazoyer, and M. Denis, "Neural Correlates of Topographic Mental Exploration: The Impact of Route versus Survey Perspective Learning," *NeuroImage*,

12 (2000): 588–600; G. Galati, E. Lobel, G. Vallar, A. Berthoz, L. Pizzamiglio, and D. Le Bihan, "A Fronto-parietal System for Computing the Egocentric Spatial Frame of Reference in Humans," *Experimental Brain Research*, 124 (2000): 281–286; G. Committeri, G. Galati, A. Paradis, L. Pizzamiglio, A. Berthoz, and D. Le Bihan, "Reference Frames for Spatial Cognition: Different Brain Areas Are Involved in Viewer-, Object-, and Landmark-Centered Judgments about Object Location," *Journal of Cognitive Neuroscience*, 16 (2004): 1517–1535.

16. Committeri et al., "Reference Frames for Spatial Cognition."

17. P. D. Martin and A. Berthoz, "Development of Spatial Firing in the Hippocampus of Young Rats," *Hippocampus*, 12 (2002): 465–480. The selectivity of "space cells" in the young rat becomes refined as the rat begins to "navigate" and not just to walk.

18. S. Lambrey, M. Amorim, S. Samson, M. Noulhiane, D. Hasboun, S. Dupont, M. Baulac, and A. Berthoz, "Distinct Visual Perspective-Taking Strategies Involve the Left and Right Medial Temporal Lobe Structures Differently," *Brain*, 131 (2008): 523–534; S. Lambrey and A. Berthoz, "Gender Differences in the Use of External Landmarks versus Spatial Representations Updated by Self-Motion," *Journal of Integrative Neuroscience*, 6 (2007): 379–401; S. Lambrey and A. Berthoz, "Combination of Conflicting Visual and Non-Visual Information for Estimating Actively Performed Body Turns in Virtual Reality," *International Journal of Psychophysiology*, 50 (2003): 101–115; S. Lambrey, I. Viaud-Delmon, and A. Berthoz, "Influence of a Sensorimotor Conflict on the Memorization of a Path Traveled in Virtual Reality," *Brain Research: Cognitive Brain Research*, 14 (2002): 177–186.

19. K. Iglói, M. Zaoui, A. Berthoz, and L. Rondi-Reig, "Sequential Egocentric Strategy Is Acquired as Early as Allocentric Strategy: Parallel Acquisition of These Two Navigation Strategies," *Hippocampus*, 19 (2009): 1199–1211; K. Iglói, C. F. Doeller, A. Berthoz, L. Rondi-Reig, and N. Burgess, "Lateralized Human Hippocampal Activity Predicts Navigation Based on Sequence or Place Memory," *Proceedings of the National Academy of Sciences of the United States of America*, 107 (2010): 14466–14471.

20. T. Hafting, M. Fyhn, S. Molden, M. Moser, and E. Moser, "Microstructure of a Spatial Map in the Entorhinal Cortex," *Nature*, 436 (2005): 801–806; K. B. Kjelstrup, "Finite Scale of Spatial Representation in the Hippocampus," *Science*, 321 (2008): 140–143.

21. Buzsáki, *Rhythms of the Brain*, 114.

22. N. Burgess, "Grid Cells and Theta as Oscillatory Interference: Theory and Predictions," *Hippocampus*, 18 (2008): 1157–1174; A. Jeewajee, C. Barry, J. O'Keefe, and N. Burgess, "Grid Cells and Theta as Oscillatory Interference: Electrophysiological Data from Freely Moving Rats," *Hippocampus*, 18 (2008): 1175–1185; T. Hafting et al., "Hippocampus Independent Precession in Entorhinal Grid Cells," *Nature*, 453 (2008): 1248–1252.

23. A. Peyrache, M. Khamassi, K. Benchenane, S. I. Wiener, and F. P. Battaglia, "Replay of Rule-Learning Related Neural Patterns in the Prefrontal Cortex during Sleep," *Nature Neuroscience*, 12 (2009): 919–926.

Chapter 11: Perceiving, Experiencing, and Imagining Space

1. My colleague Daniel Bennequin, mathematician and geometer, introduced me to this concept in mathematics. The summary presented here is his, for which I thank him; but any errors in the treatment that follows are mine alone.

2. A point has zero dimensions; a segment has one; and a triangle has two. A priori, these figures are in a plane or space, but it is also easy to imagine them in and of themselves and to move them mentally the way one physically moves a seed or a straight edge. Likewise for each natural whole number n, a simplex of dimension n is the simplest possible figure "in itself" that has the dimension n; it has $n+1$ vertices and faces in all dimensions less than n. One can imagine the points of such a simplex as the set of all the laws of probability over the set of vertices, the vertices themselves corresponding to certainties, the ridges to hesitations between the vertices, and so on, and the internal points to the case where all the summits are probable. A simplicial complex is a set of simplexes glued together.

3. In affine geometry of n dimensions, there is only one simplex; that is, two

simplexes always derive from each other by an affine (and unique) transformation. In Euclidean geometry, this is false; for example, a triangle rectangle is not isomorphic with an equilateral triangle, and so on.

4. P. H. Schoute, "Sur la réduction d'un système quelconque de forces dans l'espace *Rn* à *n* dimensions," *Archives Néerlandaises des Sciences Exactes et Naturelles*, 6 (1901): 193–196. The article was written in French, but the word *simplex* itself appears to come from German.

5. In L. E. J. Brouwer, "Zur Analysis Situs," *Mathematische Annalen*, 68 (1910): 422–434. This is done vis-à-vis the category of continuous applications and not with the much more limited one of analytic applications using pieces, as with Poincaré.

6. Alexander's article is titled "Combinatorial Analysis Situs," *Transactions*, 28 (1926): 301–329; he refers to "simplicial complexes." One might fairly consider that it marks the birth of the field of "combinatory topology," which naturally includes simplexes and simplicial complexes. The article also refers to the key notion of a singular simplex. James Alexander was a student of Oswald Veblen at Princeton. Veblen, an American of Norwegian extraction, began his career in geometry — particularly projective geometry — and then turned to classical topology and the bold inventions of the twentieth century. From 1913 onward, together with Alexander, he confirmed the results of Poincaré and extended them. Their article in the *Annals of Mathematics* of 1913, titled "Manifolds of *N* Dimensions," contained a definition of "*n*-dimensional simplex," "*n*-dimensional cell," and "complex," more or less in the sense of simplicial *n*-variety in the manner of Brouwer. In 1922 Alexander demonstrated his theory of duality in *n* dimensions that extended the two-dimensional theorem of Camille Jordan and Brouwer (and we should add Veblen to the list, for, according to Solomon Lefschetz, the first complete proof is his). Finally, in 1921 Veblen published *Analysis Situs*, which constitutes the first systematic treatment of topology. This work revisits Veblen's lectures of 1916; the 1931 edition already contains the terminology pertinent to this volume: "simplexes," and "simplicial complexes."

7. G.-G. Granger, *La Pensée de l'espace* (Paris: Odile Jacob, 1999).

8. P. Gärdenfors, *Conceptual Spaces: The Geometry of Thought* (Cambridge, MA: MIT Press, 2004).

9. On this subject, see Alexandre Koyré's remarkable book, *Du monde clos à l'univers infini* (Paris: Gallimard, 1988).

10. I. Newton, *The Mathematical Principles of Natural Philosophy*, trans. A. Motte, vol. 1 (London, 1729), 9.

11. J. C. Maxwell, "Address to the Mathematical and Physical Sections of the British Association, Liverpool, Sept. 15, 1870," *The Scientific Papers of James Clerk Maxwell*, ed. W. D. Niven (New York: Dover Publications, 1965), 220.

12. Gärdenfors, *Conceptual Spaces*. Gattis attempts to follow the path of Pinker compatible with the constructivist theories of Piaget and the ideas of Poincaré.

13. H. Poincaré, *La Science et la methode* (Paris: Payot, 1930); H. Poincaré, *La Science et l'hypothèse* (Paris: Flammarion, 1968).

14. H. Sinaceur, *Jean Cavaillès, philosophie mathématique* (Paris: PUF, 1994); J. Largeault, *L'Intuitionnisme* (Paris: PUF, 1992); J. Largeault, *Intuition et intuitionnisme* (Paris: Vrin, 1993).

15. A. Einstein, *Conceptions scientifiques* (Paris: Flammarion, 1993), 29.

16. Einstein, *Conceptions scientifiques*.

17. G. Longo et al., "Géométrie et cognition," *Revue de Synthèse*, 124 (2004): 5.

18. B. Teissier, "Géométrie et cognition; l'exemple du continu," in *Ouvrir la logique au monde: Philosophie et mathématique de l'interaction*, ed. J.-B. Joinet and S. Tronçon (Paris: Éditions Hermann, 2009).

19. K. Matoussis and S. Zeki, "Seeing Invisible Motion: A Human fMRI Study," *Current Biology*, 16 (2006): 574–579.

20. J. Piaget, *The Child's Construction of Reality*, trans. M. Cook (London: Routledge and Kegan Paul, 1954), 105–106. The philosopher Ernst Cassirer had earlier suggested the importance of the idea of the group in "The Concept of Group and the Theory of Perception," *Philosophy and Phenomenological Research*, 5 (1944): 1–3.

21. M. Merleau-Ponty, *Merleau-Ponty à la Sorbonne, résumé de cours 1949–1952* (Paris: Cynara, 1998), 190–198.

22. J. J. Gibson, "The Theory of Affordances," in Shaw and Bransford, *Perceiving, Acting, and Knowing*; J. J. Gibson, *The Senses Considered as Perceptual Systems* (New York: Greenwood, 1983).

Chapter 12: The Spatial Foundations of Rational Thought

1. On this subject, see A. Berthoz and R. Recht, eds., *Les Espaces de l'homme* (Paris: Odile Jacob, 2005), 394.

2. See my chapter in *L'Empathie*.

3. F. A. Yates, *Art of Memory* (Chicago: University of Chicago Press, 1974).

4. M. Carruthers, *Le Livre de la mémoire. Une étude de la mémoire dans la culture médiévale* (Paris: Macula, 2004).

5. J. Schied, *L'Énigme Plutarque. Quaestiones Romanae* (Paris: Vuibert, 2009).

6. D. Karadimas, *La Raison du corps. Idéologie du corps et réprésentations de l'environnement chez les Miraña d'Amazonie colombienne* (Paris: CNRS, 2005); D. Karadimas, "Chercher le centre: Stratégie d'orientation spatiale chez les Miraña d'Amazonie colombienne," in Berthoz and Recht, *Les Espaces de l'homme*, 67–92.

7. P. Descola, *Par-delà nature et culture* (Paris: Gallimard, 2005).

8. D. B. Haun, C. Rapold, J. Call, G. Janzen, and S. Levinson, "Cognitive Cladistic and Cultural Override in Hominid Spatial Cognition," *Proceedings of the National Academy of Sciences of the United States of America*, 103 (2006): 17568–17573.

9. A. Berthoz, C. Ossola, and B. Stock, *La Pluralité interprétative: Fondements historiques et cognitifs de la notion de point de vue* (Paris: Les conférences du Collège de France, 2010).

10. One might go further and ask whether humans use several fundamental concepts to represent space and manipulate points of view. This was suggested by a study of the Munduriki in the Amazon. See S. Dehaene, V. Izard, P. Pica, and E. Spelke, "Core Knowledge of Geometry in an Amazonian Indigene Group," *Science*, 311 (2006): 381–384.

11. J. Seward, *Magic Paths* (London: Octopus Publishing, 2003).

12. Seward, *Magic Paths.*

13. X. de Maistre, *Voyage autour de ma chambre* (Paris: Mille et Une Nuits, 1999).

14. A. Berque, "Écoumène: Introduction a l'étude des milieux humains (Paris: Belin, 2000), 271. See also his chapter "Lieux substantiels, milieu existentiel: l'espace écouménal," in Berthoz and Recht, *Les Espaces de l'homme.*

15. The examples are borrowed from O. Lumbroso and H. Mitterand, *Les Manuscrits et les dessins de Zola: Notes préparatoires et dessins des Rougon-Macquart,* ed. and comm. O. Lumbroso and H. Mitterand, vol. 3, *L'Invention des lieux* (Paris: Textuel, 2002).

16. Stendhal, *The Life of Henry Brulard,* trans. J. Sturrock (New York: New York Review of Books, 2002).

17. Lumbroso, *L'Invention des lieux.*

18. G. Flaubert, *Three Tales,* trans. A. J. Krailsheimer (Oxford: Oxford University Press, 1991), 71.

19. T. Landis, "Disruption of Space Perception Due to Cortical Lesions," *Spatial Vision,* 13 (2000): 179–191.

20. Michel Foucault, *Order of Things: An Archaeology of the Human Sciences* (New York: Random House, 1970).

21. Foucault, *Order of Things,* 311. The few pages of text on humans and doubles in this book contain a profound analysis of the mutations today that lead us to direct and to rethink the relationship between man and the world according to what Foucault calls the four theoretical segments: the "finitude" of man, his "empirico-transcendental," the "unthinkable," and the "origin."

22. This refers to what he calls the "obsession with space."

23. H. Bergson, *Time and Free Will: An Essay on the Immediate Data of Consciousness,* trans. F. L. Pogson (London: G. Allen and Company, 1910), 130, 133.

24. A. Berthoz, "Les Théories de Bergson sur la perception, la mémoire et le rire, ou regard des données des neurosciences," *Annales bergsonniennes,* 4 (2009): 163–178.

25. J. Bouveresse, *Le Philosophe et le Réel* (Paris: Hachette Littératures, 1998), 248.

26. J. Thuillier, *Théorie générale de l'histoire de l'art* (Paris: Odile Jacob, 2003).

27. A. Cheng, *Histoire de la pensée chinoise* (Paris: Seuil, 1997).

28. F. J. Varela, *The Embodied Mind: Cognitive Science and Human Experience* (Cambridge, MA: MIT Press, 1992).

29. J. Monod, *Chance and Necessity: An Essay on the Natural Philosophy of Modern Biology* (New York: Alfred A. Knopf, 1971).

30. A. de Libera, *La Querelle des universaux* (Paris: Seuil, 1998). According to de Libera, "Scotus's key discovery is that the human mind has the possibility to remember both its own acts and the acts of the senses . . . , this implies that it may be possible to recognize its own acts and those of the senses" (p. 325). Thus, "if the mind intuitively recognizes the sensory acts that it has achieved, it must also intuitively know the unique consequences of those acts" (p. 326). Consequently, there must be an "act of intuitive knowledge" that would itself require the presence of a "phantasm" enabling reactivation of this intuitive act. And that is none other than the action of the passage from the singular to the universal that is retained and not only the objects of knowledge in themselves. The neurologist Wilder Penfield refers to a memory or even an "experiential hallucination" that can be induced by stimulation of the temporal lobe in humans (p. 328).

31. De Libera, *La Querelle*, 328.

32. P. Alferi, *Guillaume d'Ockham le singulier* (Paris: Éditions de Minuit, 1989).

33. Cited by de Libera, *La Querelle*, 369: "Better to take fewer principles, and limited ones at that."

34. Guillaume d'Ockham, *Intuition et abstraction*, trans. David Piché (Paris: Vrin, 2005). See page 124 and subsequent pages for Ockham's analysis on the contribution of the "act of intellection," or the "*habitus*" to the mind and the relationship between the singular and the universal.

35. Alferi, *Guillaume d'Ockham*, 225.

Epilogue

1. Ricoeur, *Soi-même comme un autre.*

2. After writing the text about roofs, I found a very nice little book by T. Paquot, *Le Toit. Seuil du cosmos* (Paris: Editions Alternatives, 2003).

3. Gell-Mann, *The Quark and the Jaguar.* Murray Gell-Mann's book is a profound analysis of the quantum nature of the universe, including the workings of the brain and the emergence of consciousness.

4. As proposed by Llinás, *I of the Vortex.*

5. F. Mehran, *Traitement du trouble de la personnalité borderline. Thérapie cognitive émotionelle* (Paris: Masson, 2006).

6. See Descola, *Par-delà nature et culture.*

7. M. Edwards, *De l'émerveillement* (Paris: Fayard, 2008).

Index

Attention (*continued*)

ing, 40; feature binding theory
of, 53–54; genetic basis of, 58; imprinting, 40, 68–70; and intersubjectivity, 40; and parietal cortex,
221n30; philosophical context of,
42–45; questions on, 41; selective
attention, 40–41, 50; and selective filtering, 49–51; and *Umwelt*, 41, 46; and vigilance, 51–52;
visual attention, 221n27, 221n30.
See also Gaze

Attentional blink, 50–51, 220n19

Attneave, Fred, 67

Auditory attention, 221n30

Autobiographical memory, 37

Autopoiesis, 195

Avatars, 96, 154

Avicenna, 196

Awareness, 207

Axonal branching, 100–101, 147

Babies: attachment of, 207; and attention, 56, 57–58; control of posture and balance by, 130–31; and
emotional contagion, 57; gaze of,
30–31, 35–36, 38–39; and planar
covariation, 125; and precision of
localization, 159. *See also* Children

Babinski, Joseph, 127

Babinski's synergy, 127

Bach, J. S., xii, 17

Bacon, Francis, 212n10

Bacteria, 67–68

Baltrušaitis, Jurgis, 109

Barlow, Horace, 67

Basal ganglia, 32, 84, 103, 107, 119,
122, 123, 148

Basal nucleus, 56

Bayes, Thomas, 15

Bayesian theories, 15, 75, 105–6, 195

Baylis, Gordon C., 221–22n31

Behavior: brain structures controlling, 55; molecular basis of, 6

Bell, Thomas, 117

Bell, Tony, 233n21

Bellomo, Nicola, xi

Bennequin, Daniel, 175, 243n1

Bergson, Henri-Louis, 119, 193–94

Berkeley, George, 42

Bernstein, Nikolai, 27–28, 97

Berque, Augustin, 187

Blindness, 78

Body: celebration of, 195; expression
of emotions by, 117–19; muscular
effort of, 172–73; role of, in sensory experience, 171–76; skeleton
of, 133–37. *See also* Sensory processing; *and specific senses, such
as* Vision

Boethius, 212n9

Borderline personality disorder,
207

Boulez, Pierre, xi–xii

Bourdieu, Pierre, viii

Bouveresse, Jacques, 194

Bracketing, 13

Bradwardine, Thomas, 212n9

Brain: anticipatory mechanisms of,
85–86, 101–4, 206–7; and attention, 41, 46, 50–52, 55–57, 221n30;
and behavior, 55; as comparator,

14; and critical periods, 68–71, 91; and emotion, 55–56; as emulator of reality, 14, 53, 63–65; and evoked cortical potential, 36; executive functions of, 13, 32; and eye movements, 26–27, 30, 31–32, 104; as geometrical machine, 143–45, 176–77; hierarchies and reverse hierarchies in, 49; and identification of objects, 71–72; and inhibition, 13, 53; and inhibition of return, 49; lateralization of, 160–61, 204–5; and learning, 31–32; limbic system, 55; loops and functional networks of, 106–7; and memory, 10, 32, 65, 216n10, 248n30; and modularity, 6–7, 8; and motor command, 229n18; and multiple frames of reference and cognitive strategies, 156–58, 206; and Newton's laws, 61–62; and non-Euclidean geometries, 97, 140, 223n, 240n30; and perceptual filling, 66–67; and perceptual map, 51; probabilistic processes of, 9, 105–6; and randomness, 74–76; and reliability, 9; and saccadic substitution hypothesis, 33–35; and shape, 84–86, 87; and shortcuts by perceptual system, 62–63; and specialization, 76–80; and speed, 7, 9, 80–81, 95; and transmission delays, 95. *See also* Memory; *and specific brain structures*

Brain and Behaviour Science, 223n1

Brain lesions, 11, 38, 56, 58
Brain stem, 119
Broca's area, 76
Brouwer, L. E. J., 171, 244nn5–6
Buddha and Buddhism, 117, 195, 209
Buzsáki, György, 165–66, 168, 216n10

Camel story. *See* Weeping camel story
Carruthers, Mary, 179
Cartan, Élie, 130
Cassirer, Ernst, 245–46n20
Caste system of India, 181
Cavaillès, Jean, 174
Cerebellum, 32, 57, 95, 103–4, 119, 123, 144, 147–48, 159
Cerebral cortex, 68, 76, 85, 103, 107, 119, 160–61
Change blindness, 49
Changeux, Jean-Pierre, 17
Changing point of view, 153–54
Chaplin, Charlie, 99
Chartres Cathedral, 185–86
Chasles, Michel, 62–63
Cheng, François, 23
Children: and attention, 56–58; Piaget's developmental tasks of, 56, 69, 82, 176, 184, 245n12. *See also* Babies
Cholinergic system, 51
Chomsky, Noam, 23
Cicero, 108, 114, 235n11
Cingulate cortex, 51–52
Cioni, Giovanni, 152

Neurons (*continued*)
  plateau potential of motor neurons, 10; postsynaptic density of, 70; Purkinje neurons, 104, 147–48; receptor fields of, 79, 80, 220–21*n*25; synapse of, 75–76
Nevers attack, 99
Newton, Isaac, 172
Newtonian laws, 61–62, 81, 102–3
Noemas, 218*n*4
Noetic syntheses, 218*n*4
Noh of Kabuki, 113
Noncommutativity of rotation, 28–29, 100
Nonlinearity, 18, 104, 114
Noradrenergic system, 52

Object model verification theory, 65
Obsession with space, 247*n*22
Ockham, Guillaume d', 197, 248–49*n*34
Ockham's razor, 197
Ocular pursuit, 25–27, 30–31
Odors, 88–89, 145, 230*n*31
Olfactory bulb, 88–89
Olfactory glomeruli, 145
Olivary bodies, 147
One-third power law, 96–97
Orbitofrontal cortex, 51
OTX2, 70

Panksepp, Jaak, 117–18
Paquot, T., 249*n*2
Parietal cortex, 46, 51, 57, 58, 79, 86, 221*n*30
Parietofrontal cortical network, 156

Parietotemporal junction, 221*n*30
Paris, France, 10, 63, 135, 144, 161, 202–3
Particular versus universal, 195–97
Pashler, Harold E., 40
Pasquinelli, Barbara, 112
Penfield, Wilder, 65, 131, 248*n*30
Perception: and brain as emulator of reality, 14, 53, 63–65; of continuity of motion, 82–83; controlling uncertainty and randomness, 74–76; and critical periods, 68–71; and decision making, 89–90; and déjà-vu/déjà-vécu, 65–66; and dimensionality, 88–91; encoding the continuous by the discrete, 73–74; fragmentation of, in sensory system, 71; geometry and perceived experience, 173–75; and haptic sense, 85, 86–88; and identification of objects, 71–72, 88; Leibniz on, 44–45; link between action and, 59; little perception, 44–45; mental representations of objects produced by, 172; of movement, 96–97; phenomenal perception, 60–61; and redundancy, 67–68; resolving ambiguities, 81–83; and resonance, 66–67; and round tip of finger, 87–88; separating content from context, 83–84; and shape, 79, 84–86, 87; Shepard on, 64; shortcuts by perceptual system, 62–63; and smell, 88–89, 230*n*31; space and encoding of,